Guided Affective Imagery with Children and Adolescents

EMOTIONS, PERSONALITY, AND PSYCHOTHERAPY

Series Editors
Carroll E. Izard, *University of Delaware, Newark, Delaware*
and
Jerome L. Singer, *Yale University, New Haven, Connecticut*

HUMAN EMOTIONS
    Carroll E. Izard

THE PERSONAL EXPERIENCE OF TIME
    Bernard S. Gorman and Alden E. Wessman

THE STREAM OF CONSCIOUSNESS: Scientific Investigation into the
Flow of Human Experience
    Kenneth S. Pope and Jerome L. Singer, eds.

THE POWER OF HUMAN IMAGINATION: New Methods
in Psychotherapy
    Jerome L. Singer and Kenneth S. Pope, eds.

EMOTIONS IN PERSONALITY AND PSYCHOPATHOLOGY
    Carroll E. Izard, ed.

FREUD AND MODERN PSYCHOLOGY, Volume 1: The Emotional
Basis of Mental Illness
    Helen Block Lewis

FREUD AND MODERN PSYCHOLOGY, Volume 2: The Emotional
Basis of Human Behavior
    Helen Block Lewis

GUIDED AFFECTIVE IMAGERY WITH CHILDREN AND ADOLESCENTS
    Hanscarl Leuner, Günther Horn, and Edda Klessmann

A Continuation Order Plan is available for this series. A continuation order will bring delivery of each new volume immediately upon publication. Volumes are billed only upon actual shipment. For further information please contact the publisher.

Guided Affective Imagery with Children and Adolescents

HANSCARL LEUNER
GÜNTHER HORN
AND
EDDA KLESSMANN
University of Göttingen
Göttingen, Federal Republic of Germany

In collaboration with Inge Klemperer, Inge Sommer, and Hans-Martin Wächter

Translated by Elizabeth Lachman

Translation edited by William A. Richards

PLENUM PRESS • NEW YORK AND LONDON

Library of Congress Cataloging in Publication Data

Leuner, Hanscarl.
  Guided affective imagery with children and adolescents.

  (Emotions, personality, and psychotherapy)
  Translation of: Katathymes Bilderleben mit Kindern und Jugendlichen.
  Bibliography: p.
  Includes index.
  1. Fantasy—Therapeutic use. 2. Imagery (Psychology)—Therapeutic use. 3. Child psychotherapy. 4. Adolescent psychotherapy. I. Horn, Günther, 1935– . III. Richards, William A. IV. Title. V. Series. [DNLM: 1. Psychotherapy—In infancy and childhood. 2. Psychotherapy—In adolescence. 3. Imagination. WS 350.2 L653k]
  RJ505.F34L4813 1983   618.92'8916   83-8050
  ISBN 0-306-41232-2

This volume is a translation of the second revised edition of *Katathymes Bilderleben mit Kindern und Jugendlichen* by Hanscarl Leuner, Günther Horn, and Edda Klessmann, published in 1978 by Ernst Reinhardt Verlag. Munich and Basel. German Edition (ISBN 3-497-00821-4) © 1977 and 1978 by Ernst Reinhardt GmbH & Co., Munich

©1983 Plenum Press, New York
A Division of Plenum Publishing Corporation
233 Spring Street, New York, N.Y. 10013

All rights reserved

No part of this book may be reproduced, stored in a retrieval system, or transmitted in any form or by any means, electronic, mechanical, photocopying, microfilming, recording, or otherwise, without written permission from the Publisher

Printed in the United States of America

Foreword

A major shift in our conception of human psychology has occurred in the past 20 years. Research scientists have recognized that the stimulus–response models that characterized the dominant "objective behaviorism" of the period between 1910 and 1960 had overlooked the role of conscious thought, cognitive mapping, and related centrally generated information-processing as fundamental human characteristics. Indeed, even the psychoanalytic theorizing of that period placed great emphasis on "drives," "energy displacements," and unconscious phenomena. Today, we increasingly recognize that human beings are curious, information-gathering creatures with a differential emotional structure that is closely tied to the nature of information processing and to problems of meaning as well as to the satisfaction of physiological needs. In the basic fields of social psychology and personality theory, increasing attention is being paid to how people form constructs of the confusing ambiguities of daily life and social relationships and how they mentally play and replay these in waking or in nocturnal-sleeping fantasy form.

The clinical methods developed by Professor Leuner and his colleagues exemplify the view that human reality consists not only of the direct experiences of environmental stimulation or social interaction but also of the private reality developed as we replay these experiences in our day and night dreams and shape them into symbolic, metaphorical, or allegorical modes. The therapeutic techniques described here are designed to help individual children and adolescents confront their external reality more effectively by first leading them to a confrontation with the various ways in which they have already developed symbolic and private forms for dealing with the outside. In effect, the imagery "trips" become voyages of self-exploration that also lead ultimately to more direct, overt behavioral change. While we cannot be sure that we have a firm grasp on how some of the methods work, the elaboration of the procedures is stimulating not only for clinicians but for researchers striving to understand how we form and use mental representations of the milieu we inhabit.

This volume represents a major step—the *first* careful description in English of guided affective imagery (GAI) procedures as applied to children and adolescents, including a detailed rationale and case reports. As such, it presents a real challenge to many clinicians to explore the possibilities of this systematic and thoughtful approach in the interests of helping disturbed youth. At first, GAI may not seem to be everyone's cup of tea. It seems to some nebulous or almost mystical. And yet, the evidence from recent behavioral therapy research continues to suggest the importance of using humans' imagery capacities as part of systematic behavior change. Professor Leuner examines some of those possible links between his method and the behavior therapies to which I called attention in my book *Imagery and Daydream Methods in Psychotherapy and Behavior Modification* (New York: Academic Press, 1974). His discussion and some of the case studies presented here lead to the proposition that imagery trips not only may accomplish what behavioral treatments can do but may go even further by developing the imaginative skills of the child and thus leading to greater generalization. Research on and evaluation of clinical efforts are certainly needed to support these propositions, but the issues are now being drawn more clearly with the translation of this intriguing work.

To a generation of children growing up with television and thus strongly socialized toward a very visual, slickly packaged, and continuous storytelling experience, the GAI method may present some difficulties, but also attractions. The European children presented in this volume had a more extensive exposure to fairy tales, reading, and parental or family narration, I believe. American children may be losing the habit of reading or of private storytelling because it is so easy to become absorbed in the fast-paced American television fare. They may have some initial difficulties "getting into" a treatment such as

GAI, but I believe that once started, they will welcome the recognition of the potential richness of their imaginative capacities. With the publication of this work, we may soon find out what opportunities exist for transferring the method, apparently so successful in Europe, to helping the troubled children of America.

Yale University *Jerome L. Singer*
New Haven, Connecticut

Preface to the Second Edition

The first edition of this volume was out of print surprisingly quickly. During the past year, no fundamentally new results were gained that would justify revising or crucially expanding the manuscript. Some data were changed in the text referring to an increasingly larger number of treated cases of anorexia and to a more extended follow-up observation period in the case of "Ralph." The text was carefully examined.

Göttingen

Hanscarl Leuner
Günther Horn
Edda Klessmann

Preface to the First Edition

Since guided affective imagery first appeared in the literature in 1955, it has developed into an internationally recognized approach to the use of mental imagery in psychotherapy. It occupies a unique position among the imaginal procedures currently available to psychotherapists. On the one hand, it is distinguished by the possibility of systematically structuring the broad imaginal field of experience and, on the other hand, by its wide range of diverse technical procedures, which are flexible and can be adapted to the individual case.

The application of guided affective imagery in the treatment of children and adolescents originated in preliminary research, which Leuner conducted in the 1950s while working in the field of child psychiatry, for some years as director of a child guidance center.

The present book is the first systematic compilation of therapy with children and adolescents using guided affective imagery. The particular application for children during latency and for adolescents is summarized and critically assessed. Broader clinical experience and case examples illustrate the intense effect of guided affective imagery. In the course of treatment, the depth-psychological dynamics of the therapeutic process often unfold in a logically consistent manner. The gradual resolution of symptoms not infrequently takes place parallel to these dynamics. The concurrence of the results found by therapists working independently of one another is surprising: the modest expenditure of time with short-term therapies ranging from 4 to 25 sessions, and the successes which frequently last, without mere symptom substitution. Even with severely disturbed children and adolescents presenting considerable personality disturbances, the length of treatment is generally short, as in such therapy with adults. These experiences and observations of the release of psychodynamic processes, which, as a rule, are laden with strong affect, justify, in our opinion, the designation of *intensive psychotherapy*. The reader may form his or her own judgment about this. Further investigations with more formal research designs are planned.

The number of psychotherapists successfully using guided affective imagery with children and adolescents has recently increased. This increase has led to the conviction that the procedure has a substantial therapeutic basis by way of the depth-psychological working-through of the basic disturbance and that it produces character-changing effects. Thus, it contrasts with the recently, more frequently proposed procedures that exclude the patient's intrapsychic experience.

The authors hope the publication of their observations and results will encourage psychotherapists to verify them.

Göttingen

Hanscarl Leuner
Günther Horn
Edda Klessmann

Contents

1. The Position of Guided Affective Imagery within the Framework of Psychotherapeutic Procedures............... 1
   Hanscarl Leuner

I. FOUNDATIONS

2. Guided Affective Imagery in the Psychotherapy of Children and Adolescents............................... 9
   Hanscarl Leuner
3. Special Features of Working with Guided Affective Imagery in the Treatment of Children and Adolescents 41
   Edda Klessmann
4. Guided Affective Imagery as Used in Diagnosis in Child Guidance 59
   Günther Horn

II. THERAPEUTIC RESULTS

5. Outpatient Psychotherapy of Anorexia Nervosa Using Guided Affective Imagery............................. 77
   Edda Klessmann and Horst-Alfred Klessmann
6. The Effectiveness of Confrontation in Guided Affective Imagery in the Treatment of Childhood Phobias 95
   Inge Sommer
7. Guided Affective Imagery in Groups of Young Drug Users 105
   Edda Klessmann

8. Guided Affective Imagery in the Short-Term
   Psychotherapy of a Drug Abuser 119
   Hans-Martin Wächter and Hanscarl Leuner
9. Guided Affective Imagery in the Therapy of a Severely
   Disturbed Adolescent................................ 133
   Günther Horn
10. Guided Affective Imagery in the Treatment of an 8-Year-
    Old Neurotic Boy..................................... 159
    Günther Horn
11. Guided Affective Imagery in the Short-Term Therapy of
    an Eyelid Tic.. 171
    Inge Klemperer

References.. 183
Author Index.. 187
Subject Index... 189

The Position of Guided Affective Imagery within the Framework of Psychotherapeutic Procedures

HANSCARL LEUNER

Since the time of our first publication dealing with the treatment of children and adolescents using guided affective imagery (50), new therapeutic procedures have been introduced, which are not based on depth psychology. In child guidance centers and other therapeutic institutions for children and young people, behavior therapy and person-centered therapy are being practiced to an increasing extent today. Their significance and therapeutic efficiency, which have been only partially researched in some areas, cannot be questioned. Readers of this volume who have dealt exclusively with these procedures are perhaps not well prepared to assimilate a procedure such as guided affective imagery. For this reason, it seems important to begin this book with a short digression, as indicated by the chapter title.

As a mental imagery technique, guided affective imagery starts from two premises: (1) that human experience, even if it is only in fantasy, along with the accompanying release of affective impulses commensurate with respective individual needs, results in an intensive (i.e., fundamental) self-confrontation; and (2) that this confrontation, which is conveyed through fantasy, can be considered in accordance with the empirically gained insights of depth psychology and can best be explained by them. Not only conceptional but also technical-therapeutic consequences arise from these premises.

Guided affective imagery, which in practice has also acquired the appropriate name *symboldrama,* can therefore be assigned to that group of psychotherapeutic procedures that have elevated *immediate experience* in the interactional confrontation with itself and its surroundings to a therapeutic principle. Psychodrama and role playing can be considered its prototypes, along with play therapy for children. Other prototypes, which are more strongly verbally oriented, are imaginal narration in the stories for the Thematic Apperception Test and "elaborating on fantasies" in psychoanalysis. Play therapy has always been treated in a depth-psychological manner, as facilitating insights into the self, and more recently psychodrama has been viewed in the same way. By *depth-psychological,* we mean the recognition of an unconscious drive dynamic and counterdirected defenses, as well as the uncovering of the latent conflicts between the two. Also included is recognition of the significance of transference, countertransference, and insights into dynamic processes which manifest themselves in, among other things, symbolic representations of the dream as a regressive-magical way of striving toward resolution.

When one considers guided affective imagery from this perspective and attempts to grasp its essential therapeutically effective components, the following factors can be emphasized in abbreviated form: (1) the relaxation of general character defenses (which are effective in waking consciousness) and censorial barriers by the slightly altered state of consciousness (hypnoid); (2) the manifestation of the conflicts in symbolically cloaked fantasies; and (3) the release of suppressed impulses leading toward the tendency to satisfy archaic needs within the framework of a controlled regression (hypnoid). At the same time, the counterdirected defenses are vividly represented and are experienced in the same form. During conflict-laden trial behavior, approaches for the testing-out of new solutions are developed in the intrapsychic realm as well as in reality through the confrontation with introjected images and situations. This process is facilitated by forces of unimpeded creativity and by the innovative, imaginal depiction of future possibilities. One of the unique components of guided affective imagery that is frequently reflected in the case histories is the mobilization of creative attempts to solve problems and to make preliminary designs for future behavior.

If we consider the points of contact between guided affective imagery (GAI)—as a type of strongly experientially conveyed therapy— and behavior therapy as well as person-centered therapy and at the same time examine the differences, we can emphasize the following in simplified form. It is striking that in the comparison of GAI with *behavior therapy,* particularly in Wolpe's approach (59), relationships are present, at least in the technical aspect. In both procedures, a state of physical-psychic relaxation is used as the functional basis. In both, fantasy is mobilized so that the patient may strive for solutions through imagination and trial behavior. However, the difference between the two lies in the fact that in Wolpe's approach (a therapy that is limited essentially to the treatment of phobias) desensitization is practiced by confronting anxiety-laden situations from real life. The gradual structuring of the anxiety-laden events ("hierarchy") takes place in a state of full waking consciousness, that is, before the patient has entered the relaxed state of altered consciousness. In the second step, these objects are imagined in the state of relaxation and the anxieties are extinguished by progressive confrontation. Thus, the concept of behavior therapy proceeds on the assumption that the neurosis can be equated with its symptoms. Intrapsychic processes and their perception have only marginal significance. The concept of behavior therapy denies the relevance of symbolic processes and, accordingly, must face their appearance in the imagination uncomprehendingly. For this same reason, the behavior therapist is not acquainted with working directly on symbolic manifestations, work that is extremely productive in GAI and shortens therapy. As condensations of broad conflictual experiences, symbolic manifestations aim at the core conflict. In dealing with those symbolic manifestations in the management model of symbolic drama (for example, by confronting the symbol) therapies can be dramatically shortened, as is expressed by Sommer in Chapter 6. In addition, working with GAI has shown that even without constructing a hierarchy of anxieties, patients in spontaneous imagery are quickly capable of completing a hierarchically graduated, symbolic mastery of anxiety-producing situations in an astonishing, as it were, preconscious assessment of their capacity to withstand stress. They can do this in approximately five sessions and thereby achieve the lasting cure for a phobia, which has existed for many years. The basic idea behind GAI therapy is that "under their own direction," patients can move along their therapeutic paths, freeing themselves from symptoms by using merely what the symbolic-dramatic development makes possible and with just a little guidance by the therapist. This basic idea finds convincing support in observations of the following kind. As an example, I can refer to the treatment of an 11½-year-old girl with a bridge phobia (p. 31). The impression has arisen that GAI is superior to behavior therapy in many respects not only because of the remarkable reduction in treatment time but also because the therapy covers broad areas of the personality, and because diverse symptoms, apparently not related to each other, can be resolved through this process. This impression is evidently the result of the broader solution to deep-seated background conflicts in accordance with the concepts of depth psychology. These observations are further confirmed by the results of the so-called "operation on the symbol" (7).

At first, the points of contact of GAI with *person-centered therapy* (formerly called *client-centered*) as practiced by Rogers (45) and Tausch (53) seem to be negligible, since this therapy occurs primarily on the verbal level and neither reflects nor approaches the procedural area of experiential therapy. However, the experiential part of person-centered therapy is located in the therapist's desired variables. They are defined as:

1. Positive regard and warmth
2. Genuineness and self-congruence
3. Active participation
4. Verbalization of emotional contents
5. Nondirectiveness

The therapist's adherence to these variables constitutes a particular type of interaction, which can also be successfully transferred to GAI and characterizes the "emotional climate," in which GAI therapy also can develop favorably (7).

Seen from this aspect, the person-centered therapist will likewise be able to gain access to GAI. By participating in empirical introductory seminars in GAI, he will soon recognize that the behavior typical of person-centered therapists requires modification here. The imaginal contents and their exploration take precedence over the basic verbal behavior of person-centered therapy. However, the insight that the result of therapy is substantially determined by the behavior of the therapist is common to both procedures.

I

FOUNDATIONS

2

Guided Affective Imagery in the Psychotherapy of Children and Adolescents*

HANSCARL LEUNER

*Expanded version of the article as it first appeared in *Praxis der Kinderpsychologie und Kinderpsychiatrie, 19* (1970) 212.

INTRODUCTION

Psychotherapeutic procedures, which are directly addressed to children and adolescents, place their methodological emphases on play techniques and on verbal exchange. With infants and also with schoolchildren, it is specifically play—for instance, in the form of the scenotest reworked as the "scenodrama"—that provides diverse information about unconscious conflicts, fears, and neurotic attitudes. Frequently, the milieu responsible for the child's upbringing is also reflected in the perspective of the child. In play therapy, which can be freely developed from play, the child is able to find solutions, and a therapeutic process specific to the child unfolds without verbal interpretations being given. However, there are limits to play therapy, on the one hand, in regard to the age of the child and, on the other hand, in relation to particular psychic structures. Compared with play therapy, verbal exchange, which is more appropriate for the older child and the adolescent, remains for the most part limited to the exchange of factual information and of orienting opinions. The emotional involvement inherent in play is lost. Information about depth material remains scanty for the most part. As a rule children of this age succeed in concealing preconscious anxieties and conflicts in accordance with their defensive processes, which are already more strongly developed. Revealing these anxieties and conflicts is no longer compatible with the values of this age group. Working with dreams or free association is also usually unproductive with this interim group, those between approximately 8 years of age and the end of puberty. These difficulties are even more accentuated in schizoidly and overly intellectualized children. Sensitivity and extremely delicate sensitiveness are often hidden behind these children's character armor. Just at this point, GAI has proved to be successful as a continuation of productive play therapy using other means. With the help of GAI the patient gains renewed access to experiential areas withheld from consciousness through bypassing the commonly found and undesired defensive processes described.

The Nature of Guided Affective Imagery

Initially we developed GAI as a method for experimentally verifying unconscious processes. Hence, we gave it the scientific designation *experimental guided affective imagery* (EkB; in German-speaking countries, the procedure was introduced as *katathymes Bilderleben*, or *catathymic imagery*; therefore, *Experimentelles katathymes Bilderleben* = EkB) (29). In therapeutic practice it was subsequently labeled *symboldrama*. The technical origins of the procedure are to be found in the unstructured methods of Kretschmer's (26) "filmstrip thinking" and Happich's (19) instructions, which were later developed into religious meditative exercises. In the past 20 years many articles about GAI have been published. They deal with the experimental as well as with the therapeutic aspects (27–40). As far as the technical aspect is concerned, a series of therapeutic interventions was developed. Through this, methodological approaches were created out of the immeasurable breadth of the mental imagery experience and order was brought to its diversity. At the same time GAI became teachable as a psychotherapeutic method that was technically relatively easy to master. It proved to be successful particularly as a form of short-term therapy for adults (36). Symboldrama seems to achieve even more favorable results with children and adolescents. The greater plasticity of the child's psychic life and his genuine readiness for preconscious imaginal productions seem to be responsible for this. GAI complies with the requirements of child therapy for another reason. Psychotherapy with children is aimed not only at the treatment of the child himself but must also reach the parents with regard to their principles of upbringing. Child therapists who have used GAI have shown that through its aid, it is easy to get across to parents insights into their incorrect pedagogic stance. These therapists have let parents know passages from the imagery produced by their children and have then explained to them how their child emotionally experiences the family milieu. The prerequisites are, of course, parents who are suited to this approach and psychotherapeutic tactfulness.

As already mentioned symboldrama is based on using fantasy production and mental imagery therapeutically. Both categories have been known to psychoanalysis for a long time. In psychological terms, the kind of mental imagery we use is known as *hypnagogic vision*.

Psychologically speaking, GAI is a projective procedure. The object is projected into the darkness "in front of the closed eyes," and the means of the projection itself is the optical fantasy. In contrast to all of the known projective procedures (above all, the well-known tests) GAI is distinguished by its independence from given material structures. Consequently each slight change in the intrapsychic constellation is immediately reflected in the envisioned images, a process that we call *mobile projection* (29).

Symboldrama differs from the earlier unstructured methods—for example, from Kretschmer's (26) "filmstrip thinking"—in an additional way. Instead of waiting for spontaneous imaginal productions, basic imaginal motifs are suggested by the therapist, for example, the idea of a meadow, a house, or a mountain. In the state of deep relaxation, these ideas are manifested in the form of lively, imagined, colorful pictures. Their individual design reflects in a symbolic way the dominant neurotic problem-situation of the person concerned. Through this technique of GAI production, one can relatively reliably count on stimulating psychoanalytically relevant thematic cycles into visional representations.

These *imaginal motifs,* which the therapist suggests to the patient, are analogous to the image charts of the projective test, for example, of the Thematic Apperception Test. In this way the respective themes, which the psyche comments on projectively, are brought up both for diagnosis and for further therapy. Later we will discuss which of the 10 standard motifs we have developed are preferred in the treatment of children.

As a directive mental-imagery technique, GAI has still further possibilities; namely, the therapist can intervene directly in the flow of the images. In so doing, he can direct the symbolic processes and, in addition, can encourage a clarifying dialogue at the precise moment when the affective image is envisioned. Moreover, the associative field can be extended into the past, as well as taking into account the here-and-now of the transference and of the current life situation.

In comparison to the *psychoanalytic technique*, therefore, our procedure unites the temporally separate events of the dream, on the one hand, and of the free association with the recapitulation of the dream, on the other hand, to form a single process in the presence of the therapist. The patient dreams, so to speak, under the protection of the therapist, who directly participates in the dream experience through his communications. The patient is just as emotionally involved in this dream as in a night dream but under the therapist's protection is also capable of really dreaming the dream to its completion. This procedure has consequences for the therapeutic process and requires a treatment technique that is different from, but by no means simpler than, psychoanalysis. One can imagine that this process is emotionally laden and requires a fluctuation across various levels with regard to the state of consciousness, similar to that which Bellak (6) postulated for free associations.

When considered from the standpoint of the therapeutic dialogue, and thus of the relationship of transference and countertransference, GAI is characterized by the fact that the interpersonal emotional field is changed by the "inserted" level of imaginal projection. It is the actual and central spot where the therapeutic process is reflected and becomes visible to the patient and the therapist. The contents of the imagery often have the character of impressive experiences. They relatively rarely require extensive or indepth interpretation. Unconscious problems, repressed anxieties, wishes, and expectations, along with the diverse defensive processes and neurotic character resistances, are demonstrated to the patient. The contents of GAI can frequently be interpreted as a mirror of the unconscious psychic life itself and can encourage the patient to face new confrontations. Despite these differences in technique, in its theoretical foundations GAI must remain bound to psychoanalytic theory if it is to be evaluated productively. It is a psychoanalytically oriented, short-term psychotherapy.

From the standpoint of technical implementation GAI has an additional advantage. In order to practice it, one does not absolutely need to have completed the entire psychoanalytical training. The therapist who has completed further training in depth psychology, for example, to the extent that is currently required in Switzerland for professional psychiatric training, can employ GAI after appropriate training in the procedure.* For didactic reasons, and to do justice to the conditions in Germany, we have divided symboldrama into three levels (elementary, intermediate, and advanced). Thus, it is adapted by steps to the technical prerequisites and the theoretical preparatory training of the therapist. The following presentation does not expressly take this gradual division into consideration; it corresponds approximately to an intermediate level.

The Technical Implementation

The technical implementation of GAI is relatively simple. While lying on a couch or sitting in a comfortable armchair, the child is encouraged to relax. For adults, the first two stages of autogenic training are helpful. As a rule, simple suggestions for resting and relaxing are sufficient. For many children even these are superfluous. It suffices to ask the child to lie down comfortably and to close his eyes. Of course it is necessary to have made supportive contact in advance in one or more sessions and to have taken an anamnesis. According to the behavior shown, easily recognizable by the amount of eyelid movement or by the kind of breathing, one may still find remains of tension or inner restlessness. One can then add a few relaxation suggestions. Images of a calming situation can be successfully awakened; for example, one is lying in a green meadow in the summer's sunshine and is resting after a long walk. The next step is conducive to the manifestation of GAI. One asks the child to picture one of the affective images that are listed in this chapter. As a rule, we begin with the meadow motif. We do not suggest a particular meadow or one

---

*The local branch "Therapy of Children and Adolescents" of the study group for GAI and imaginal procedures in psychotherapy (AGKB) regularly organizes seminars for attaining qualification as a "GAI therapist" in this field. (Headquarters: Von-Siebold-Straβe 5, 34 Göttingen, Federal Republic of Germany.)

arranged in a certain way, but just "a meadow." It is irrelevant whether the meadow exists in reality or if it is an imaginary meadow.

In a state of relaxation, images that are generally only vaguely crystallized take shape more plastically, more vividly, and more eventfully than in a state of clear waking consciousness.

It is unnecessary to go into detail regarding the following points. The child soon slips into a *light hypnoid state.* Theoretically, all of the positions are present, as Schultz (47) described them for autogenic training under the concept of *shifting over*. Thus, within a short time, the patient "sees" the meadow so vividly that he seems literally to stand there and can soon freely move around in this imagined landscape, as if it were "another world," perceivable with all of his sense organs.

The therapist supports the visual character of the affective image by acting, through comments and questions as if it were, in fact, a "quasi-normal" world. The patient is asked to report continually on what he observes within the catathymic panorama. In this way, close verbal communication is kept up during the entire therapeutic process.

The images that we have induced in the child are no longer simply the real objects but have taken on the character of a depth-psychological symbol on this level of imaginal consciousness (a concept coined by Happich 19, and Heiss, 20). A cow envisioned in the meadow, for example, can be the symbol of the mother image, an elephant that of the father image, and so forth. It would take too much space to go into detail here regarding the problems of symbolism within this framework. Let it suffice to refer to Silberer's (49) phenomenon of "autosymbolism," which led to an expansion on the Freudian concept of symbol (15). Silberer was able to show that hypnagogic visions are the expression of an internal psychic condition. Frieda Fromm-Reichmann (16) extended the significance of the imaginal productions with regard to daydreaming. She advocated that the element of unreality should not be overestimated in daydreams, as psychoanalysis has previously done. As psychic processes of the private realm, daydreams doubtless reveal, on the one hand, wishful fantasies; on the other hand, in so doing they always refer to the actual problems of the patient. Thus, daydreams need not necessarily represent an escape backward. Rather, the daydreamer succeeds in objectifying his conflicts in imaginal figures. This aspect is of central importance to our procedure, as will become clear when we turn to the diagnostic possibilities of GAI in the following section.

THE STANDARD MOTIFS: THE DIAGNOSTIC ASPECT

On this level of mental imaginal consciousness, all sorts of phenomena produced by fantasy are possible. This is the case to an even greater extent in our technique, as this level becomes intensified to a dreamlike state. In his spontaneous method of "filmstrip thinking," Kretschmer (26) was already able to show that asyntactical associations can be observed with the typical signs of Freudian dreamwork, displacement and condensation (15). In a more directive and inducible procedure (as in GAI), however, above all, landscape motifs (32) and distinct scenic associations are formed. From the immense diversity of possible manifestations we chose those for our procedures, which were most frequently produced spontaneously and are, at the same time, psychodynamically relevant. They have proved to be particularly successful in the course of long clinical testing with patients.

The first standard motif is a *meadow*.

Symbolically speaking, the meadow motif has broad significance. It can be seen as a new beginning. It may contain the quality of a cheerful, carefree holiday in summer. It can have the attributes of fertility (in the sense of generous motherliness) and of the unexpected encounter. It can represent a vast encounter with nature; flowers, plants, and a brook; the edge of the woods and the woods themselves, with the main path; the nearness of a mountain; nature in the form of weather and of the seasons; all kinds of animals; and even working and celebrating humans. Therefore, as a rule all of our exercises begin with the meadow, as long as other motifs do not intrude spontaneously, and end again with the meadow. Therefore, we also have the patient describe the meadow including its boundaries, the weather, and the more distant surroundings. An imaginative and emotionally suggestible child will develop an abundance of projective contents, which provide important references to the unconscious emotional life and lead to therapy in the associative procedure.

The following diagnostic *examples* may elucidate some projective reflections through the standard motif of the meadow.

An 8-year-old girl who had been limited and intimidated by a narrow-minded family upbringing pictured herself in a meadow. However, she stood in an enclosed section and could not find a door leading out. The grass was trampled upon, the weather gray, and the atmosphere altogether gloomy. No human being nor other form of life was in the vicinity. This sense of abandonment and of limitation seemed to be a reflection of her own mood and also expressed the subjective experience of her own life in general.

Occasionally, adolescents see endlessly expansive meadows, over which their eyes can wander for miles. This view can also express a type of abandonment, although, of course, of an entirely different sort. The lack of limitation and the seemingly endless expanse are often connected with a certain perplexity about where to go, and they give indications of illusory expectations regarding the patient's future life.

Meeting animals in the meadow, whether they appear spontaneously or are induced by the therapist, permits an examination of emotional relationships to figures in the child's personal environment in symbolic form. The cow, which is often found in the meadow, can serve as a means of learning about the relationship to the mother. For instance, a child may become anxious when we suggest that he approach the cow. The cow may turn away or point its horns at the child in an unfriendly manner. It is also significant whether the cows in the meadow are well fed and have shiny coats or stand there dirty and emaciated. If a bull that is grazing in an adjoining meadow furiously stamps the ground as an adolescent patient approaches the cow, we can draw conclusions about the child's Oedipal situation. Shy deer that come out of the woods onto the meadow may flee anxiously. From this reaction, we can suspect that the child himself is shy and, under the slightest amount of stress, seeks refuge in the certainty of psychic withdrawal, a frequent reaction of sensitive, anxious children.

Observing the brook in the meadow also provides us with a great deal of information. A small, dry rill is a sign of the suppression of vital, psychic drives. In contrast, we most often find a briskly bubbling, clear brook with lots of water in children who are either slightly neurotic or healthy. However, if, in tracing the further course of the brook, it soon trickles away or dams up, or its waters become muddy, we consider these distinct signs of disturbance, either acute or chronic.

The diversity of these possibilities stemming from the meadow motif points to a number of further standard motifs, which developed naturally from the meadow and have been used for GAI. From the total of 10 main motifs and 2 auxiliary motifs, we ordinarily use only the following 8 in therapy with children:

1. The *meadow*, as the starting point of every therapeutic session.
2. The *ascent of a mountain*, in order to obtain a panoramic view of the landscape from its summit.
3. The *pursuit of the course of a brook* to its source and/or downstream.
4. *Visiting a house*, which the child enters in order to explore its interior from top to bottom.
5. An *encounter with relatives*, as real figures or symbolically disguised as animals, so that we can gain insight into the child's relationship to parents and siblings or to "authority figures."
6. The *observation of the edge of the woods from the meadow*, so that figures emerge from the darkness or enter the woods.
7. A *boat* that suddenly appears on the shore of a pond or of a lake. The child then climbs on board and either is a passenger or steers the boat himself.
8. A *cave*, which we first observe from a distance to see if symbolic figures emerge. Then the child enters, if he so desires, either to stop for a while or to explore the cave's depths.

We now present a few examples of the subjective elaboration of these motifs to give the reader insight into the symbolic manifestations of GAI.

Ascent of the Mountain

This motif provides information about a variety of areas. Investigations by Kornadt (25) have shown that the height of a mountain statistically correlates with the patient's level of aspiration, as does the height of towers, houses, and all high buildings (the scenotest is also similar). Ambitious people and those who have high illusory expectations regarding their own achievements and capabilities see extremely high mountains; those who are discouraged and depressive, on the other hand, frequently see low mountains. The ascent can additionally be viewed as the subjective perception of one's own achievement potential and of the confrontation with mastering life and the world. Wishful thinking is expressed when the patient suddenly reaches the summit, as if by magic, without having exerted himself.

A discouraged stutterer climbed a particularly high mountain with a great deal of exertion. He slipped on muddy paths again and again. When he had finally reached the peak, a heavy snowstorm began and he clung anxiously to the summit cross.

The mountain can also be interpreted on the object level, where it frequently symbolizes rivalry with the father. Another adolescent stutterer first climbed a relatively steep but not overly high mountain. From the top, he sighted another longer and higher mountain beyond a valley. On the summit of this mountain stood an observation tower. Inspired to climb this other, "mightier" mountain, he succeeded only with effort. When he had reached the platform of the observation tower, a thunderstorm came up and lightning flashed. In accordance with the total constellation of this patient, it was clear that he had attempted an identification with his father through climbing the mightier mountain and its tower, a climb that caused fear of punitive anger. The problem was subsequently worked through. Residual symptoms of the child's stuttering, which were manifested particularly in the presence of his father and other authority figures, then stopped.

The panoramic view of the landscape is similarly revealing. Not only the dimensions ahead (the future), behind (the past), and to the right and left may be significant. It is, for instance, a characteristic of unrealistic, illusory expectations, if a young girl regularly finds spherical clouds beneath her without ever perceiving earth. At the beginning of treatment with adolescents, the landscape is frequently as it appears in early spring. Toward the end, it tends toward the ripe fullness of summer. Initially, the landscape is little frequented or acted on by people; only wide fields, a desert, or a rough mountain landscape is visible. In the course of treatment the landscape develops into one familiar in our society, in which people live, with villages, fields, and traveled streets. In the distance are cities and industrial plants, further signs of human activity.

Pursuit of the Course of a Brook

As mentioned above, obstructive motifs, such as the trickling away of the brook or an obstacle in its course, are frequently symptoms of neurosis. Besides the significance of the brook as an expression of vital drives, it can also represent a boundary, as will later be shown in the case of a girl with a bridge phobia. The brook motif is suited therapeutically for symbolically refreshing oneself. One can wade in the water, one can even swim in it, and above all, one can move on to its source and drink the fresh water there or cool and massage painful or otherwise ill parts of the body. Invigoration and compensatory comfort can be a part of bathing in the banked-up source (the origin, the giving element) and also of bathing in the sea. In some cases of therapy with adults, we have observed an unexpected improvement in psychosomatic ailments through these measures.

The House

Freud (15) saw a symbol of the personality in the house. This relationship is particularly pronounced in adults. On the other hand, children and immature personalities see in the house their own parental house. Imagining the house provides useful information about the family life of the child and also of the adolescent, whether the house is a largely realistic reproduction (as in the scenotest), or whether a "dream house" appears. The house and its rooms, its furnishings, and so forth reflect an abundance of personal or family problems and, often as well, wishful fantasies within the family sphere. However, the Freudian aspect is also obvious in adolescents as our examples will indicate.

Technically, we indicate to the patient that he will pass by a house on the way through the countryside. As the house is approached from the outside it is already significant whether a young girl pictures a baroque castle with ladies-in-waiting in its park or whether an 11-year-old boy who is extremely inhibited and reserved sees a relatively simple house without windows and doors.

In the following example it becomes clear how the house changes according to changes in the internal psychic situation. In various sessions a young girl repeatedly pictured either office buildings or restaurants. They were devoid of any personal, friendly atmosphere and were, on the contrary, businesslike and gloomy. However, after the patient had suddenly fallen in love, instead of these buildings she saw an idyllic forester's house in the woods with a garden full of luxuriant flowers, ripe pumpkins, and cucumbers. Inside the house, she found a library decorated with hunting weapons and trophies.

The important locations in looking through the house are the kitchen (bringing up the oral sphere) and the living room as an expression of comfortableness, of atmospheric reflection (to be assigned to the anal sphere, if the catch phrase "warm, bad air" is brought up). The contents of the bedroom and its closets, which may be looked through, help in assessing the development of eroticism and sexuality and also the bond to the parents. Closets, chests, and trunks in the attic and in the cellar may also be looked through for memories from childhood; old toys, family albums, and picture books.

Encounter with Relatives

This encounter may also take place in the meadow. The patient is told that he will see a cow (representing the mother) or an elephant (representing the father) coming toward him from the distance. One may also suggest that a person will approach the patient from the distance and, after this vision has been realized, imply that the figure will turn out to be the father, the mother, or some other figure (for instance, a teacher or a boss). We ask the patient to observe the figure. We may even urge the patient to hide behind a bush for his own protection. We may also ask the patient questions about his feelings toward this figure. The patient may finally disclose the figure's identity and then report about the figure's behavior toward him. Very different forms of approach, as previously with the cow, provide the therapist and the patient with information. Disguising important relatives in animal symbols has the advantage of avoiding the provocation of defenses.

A very revealing example of two brothers may illustrate this problem and also point out the diagnostic value.

Two brothers, Erwin, 11 years old, a delicate asthenic type, and Heinz, 10 years old, just as big as his "big" brother, but more robust and more self-confident, experience their relationship to their strict and irascible father in different ways:

Erwin is instructed to hide behind a bush in the meadow, because an elephant will be trotting along. The elephant appears treading stormily along with his trunk raised and trumpeting. He discovers the boy and heads straight for him. The boy flees and runs to a village. The elephant follows him there, too. Finally, Erwin succeeds in running away to an old farmhouse. He locks the door with a key. But the elephant demolishes the top part of the door with his trunk and pokes it into the house in order to grab Erwin. The scene is clearly anxiety-laden; no solution is found.

The situation is different for the younger Heinz. As the elephant comes trotting along, the landscape is transformed into an African jungle. Blacks carrying spears come out of ambush and drive the elephant into a readied pen. Here they subdue the raging animal with their spears and tie its legs with ropes so that it comes under their power.

The varying amount of anxiety directed toward the irascible father (i.e., the danger emanating from him) is averted in very different ways. The older but asthenic boy helplessly searches for protection and is pursued anyway. The more robust, younger boy dares to confront the danger actively. These catathymic contents gave rise to extended discussions with the young patients about analogous familial scenes and the way they mastered these individually. Extensive child guidance work followed.

Not infrequently, the entire family constellation is displaced to the level of animals. Thus, for instance, a 9-year-old boy pictured a large badger that lived with three young ones in a den. The patient spontaneously identified the badger with his mother. Only after a while did it occur to him that a father could be there, too. He finally appeared in the underbrush making considerable noise and was dragging his prey. All the family members pounced on it. The actual situation in the family was that the very busy father was frequently "invisible" in the family, although he dominated to a considerable extent. His generous and supporting attributes were placed in the foreground as in the described scene (29).

Observation of the Edge of the Woods

The goal is to get the patient to picture relatives or symbolic figures that personify unconscious anxieties or personal aspirations. The symbol of the woods is relatively diverse for the child, in contrast to the images of the adult. On the one hand, it can be interpreted as a symbol of the unconscious with qualities of security or of threatening danger. Repressed material develops from it; that is, symbolic figures appear, which then present themselves in the open meadow. After waiting patiently while watching the dark woods from the standpoint of the meadow, figures begin to appear. One can also designate these figures in advance, if particular ones are desired. This method has proved particularly useful with adolescents. Children, on the other hand, experience the woods as providing shelter, as concealing and as giving protection. In the meadow, however, they may feel "abandoned." What the wood has to offer may therefore be very different for the younger and the older patient and can give important clues to the unconscious experience. Animals, such as a shy hare or a deer, as well as tramps and hunters and occasionally even a witch may emerge from the woods. Younger children, particularly, may also wish to enter the protecting woods. In this case, the woods also seems to have maternal qualities, in the sense of protecting and guarding, as if the child would like to "crawl under her skirt" for his own protection. Regressive tendencies can be seen there and may become effective in the sense of therapeutic regression (Balint, 2).

For patients in later puberty, a stag presenting himself nobly, a sluggish or lazy bear, a shy, good-natured giant, and the like can appear. For children, on the other hand, fairy-tale-like figures are naturally more frequent. Such fairy-tale contents may possibly suggest that the therapy proceed on this "fairy-tale level." For children, this is an "adequate way of conversing." With increasing age, however, we would perceive such infantile-regressive characteristics as a defense against reality or as a tendency toward illusory wish-fulfillment. We would not encourage it for any great length of time but would introduce the real aspects to the patient, taking age into account. Cautiously interpretive references, which reflect the individual behavior, and comparisons with real situations can serve reality control, although persistent adherence to reality can also emerge as a form of resistance. Productive regression to the sphere of fairy-tale-like fantasy can be therapeutically desirable to encourage young patients toward creative emotional naturalness.

Children who persistently stick to realistic contents can be instructed to invent their own fairy tales.

In contrast to Desoille's procedure (10), in GAI we avoid, if at all possible, *offering* solutions to tasks through magical or fairy-tale-like practices. We do not allow the patient to change the world with a magic wand in order to appease symbolic figures that produce anxiety, or to give himself magic powers.

We are of the opinion that such imaginary tendencies are not suited to encouraging the development of a more mature ego but that they merely suggest magical and omnipotent solutions.

An Example of the Boat Motif

The case of the 8-year-old stutterer Heinz illustrates the meaning of the boat motif, which also has considerable diagnostic significance. Both of the child's parents are teachers, "nice" but ambitious. To the child, the father is the admired and unattainable model, whom he constantly overactively emulates, only to be repeatedly disappointed because he does not succeed. Such was the case in a swimming course, which the father took along with him. From time to time, Heinz fearlessly put his head under water and made a few quick swimming movements but could not let himself be calmly supported by the water. In school, Heinz always wanted to be first. The grade of "B" was considered pretty bad; a "C" was a catastrophe.

He was the oldest child in the family and had two sisters.

When the boat motif was introduced, Heinz immediately saw himself in a small paddle boat in the middle of the Baltic Sea. A friend, who was his intellectual inferior, sat behind him. For a short time he could elatedly enjoy the boat peacefully gliding along, following his command. However, a thunderstorm soon came up. He lost an oar and drifted helplessly in the sea. Waves broke against the boat, and the gliding of the boat was, as it were, just as abruptly interrupted as the flow of his speech when he stuttered. In this image he then spontaneously detected his father, who stood on a bridge, plunged into the water, and brought the boat safely to land. (The feeling of being exposed to danger on the "high seas," the fear of punitive authorities, and the dependency on the father as the "rescuing angel" become obvious.)

An Example of the Cave Motif

The emotional significance of the cave motif for children becomes evident in the following example:

A 7-year-old boy with a severe anxiety neurosis did not feel at ease in the meadow. He enthusiastically made his way into some "thick shrubbery," which he then perceived to be a cave.

("What is it like there?")—"Nice and dark, nobody can see me."—("But you can see everything outside?")—"Yes!"

He felt so safe and courageous in his hiding place that, to the therapist's surprise, he suddenly burst out and grabbed a hare, which he was supposed to watch, and "threw it far away . . . then it was dead." Then an unexpected reaction, in the sense of a self-punitive tendency, occurred: A thunderstorm came up; a previously warm spring had suddenly become cold, and so on.

The sheltering and concealing character of caves and also of hollows made of leaves in the woods is repeatedly apparent with younger children. Not infrequently, these caves have strong libidinal overtones. The 8-year-old stutterer (previously mentioned) took his sister, who was two years younger, into the cave when he was asked whom he would like to invite into the cave. He made "comfortable beds out of leaves." Then, he got himself something to eat and felt safe from the view of adults. The "cuddly intimacy," seeking protection from adults, is frequent in this age group.

For a 17-year-old girl with pronounced sexual anxieties, the cave became the spot to which she transferred her specific anxieties.

"Branches and brushwood hang down over the entrance to the cave from above, so that one can scarcely look in. There is a gully, which goes deep into the cave. Inside, it is moist, damp leaves . . . the cave is exactly as big as I am . . . I'm afraid that there could be disgusting little animals . . . " After a while a "disgusting, fat, greasy, yellow caterpillar" wanted to push its way into her cave. When she was instructed to feed and to confront the glances of the animal and finally to stroke it in reconciliation, an interesting transformation occurred. The caterpillar became "really cute, soft, with little black hairs." The phallic aspect of sexuality seemed to have been transformed into a more tender, libidinal aspect. The patient perceived the change as decidedly enchanting. After this session she had intimate relations with her boyfriend, without being afraid for the first time.

The woods also quite often evokes a similar intimate experiential sphere with feelings of safety and concealment. Trees are then perceived as protectors, particularly if they have leafy crowns. With adults, however, the woods more frequently have the character of the "uncanny unconscious," in the opposite sense, so that they are clearly avoided.

At this point, we insert a case from therapy, which illustrates the above particularly well. (In place of the edge-of-the-woods motif, we chose the cave motif, which does not involve a fundamental difference.)

After emigrating from East Germany, a 17-year-old high-school student, Hans, became ill, showing the symptoms of an autistic puberty crisis. The possibility of a beginning hebephrenic schizophrenia could not be ruled out. Taciturn and suffering from contact problems, he spoke haltingly and with a rigid face. He also complained about frequent headaches, which could not be organically explained. Before the resettlement, he had been a rather gifted student. In the new school he failed completely. The symptoms had begun four months previously, triggered off by his being in a resettlement camp.

In GAI, we again started with the meadow. A cave spontaneously appeared at the foot of the mountain. The patient was urged to observe the entrance to the cave. After a while, a fierce giant appeared. However, he stopped timidly at the exit from the cave and finally motioned to the patient. Encouraged to follow the giant, the patient was led into the cave and shown his world: a landscape with a river, vineyards, storerooms, and a barn full of cows, almost a small, luxuriant paradise. The patient was encouraged to ask the giant his reasons for having withdrawn from the world. He had fled there because he had been avoided and laughed at.

In the three subsequent sessions of GAI, each lasting three-quarters of an hour, the patient was repeatedly urged to lead the giant back out of the cave and into the world. Here, he encountered farmers who were working in the fields. (This corresponds to the technique of the training procedure.) He was instructed to roam through villages and cities. In the course of these wanderings, the giant's stature grew gradually smaller. Having shrunk to the size of a child, he at one point lay down in the bed that the patient had formerly slept in as a child. We let him rest there. Soon, he again appeared, all the more energetic, among the working farmers. Having become more cheerful and trusting in the meantime, on our advice he took a job as a houseboy in a hotel that had magically appeared not far from the cave. At the same time a synchronous transformation took place. Two rock ledges, which had previously constricted the river near the cave and had dammed up its water, had now been separated making room for the stream and a newly constructed, wide highway. Such synchronous transformations give indications of the progress of therapy (6). At the same point, the patient's symptoms had improved substantially. He was not only more talkative, affectively more relaxed, and more flexible but also within a very short time became acquainted with his new classmates for the first time. Soon thereafter he participated in sports competitions and brought home good grades. Meanwhile, the headaches had subsided to a tolerable level.

The motif of autistic seclusion and of unrealistic expectations stands out distinctly, combined with wishes for omnipotence (the giant) and with the rejection of competitive confrontation with the real world. It was astonishing to see the transformation on the level of the catathymic fantasies running parallel to the transformation in outward behavior. This is not ordinarily the case. Frequently the outward adaption lags behind. Besides the therapeutic effect brought about through purely a training procedure with the range of realistic situations, the diagnostic effect was also interesting. It showed how situationally produced competitive situations can be worked through emotionally in the labile condition of puberty. The rapid therapeutic success was surprising to us. The situational neurotic condition was, of course, only compensated for through this brief intervention. Approximately one year later, the patient was presented to us again with less severe symptoms.

It is advisable to add a few comments regarding whether and to what extent patients "understand" the symbolic connections they have pictured. As a rule, merely through the easily comprehensible contents of GAI, patients gain a certain kind of "insight," as was doubtlessly true in the above case. For this reason, the procedure, described as predominantly diagnostic, cannot really be separated from the therapeutic aspect. This lack of distinction has created complications—for instance, in the attempt to expand the method into a projective test. Every experience of catathymic imagery stimulates internal confrontations, and even after the first diagnostic attempt, the patient is no longer quite the same person as before. This internal process of change can occur preconsciously for long periods of time, as in the case just cited. The insights gained take place more on the level of emotional insights (34). However, the patient is not always in a position to recognize the relationships rationally and completely, for example, between the symbolic figures and real figures or between them and his own behavior. That recognition usually only occurs in a process of gradual clarification, in which the problem is touched on again and again in GAI. In this way, it gradually loses affective strength, moves increasingly closer to consciousness and gains access to more mature forms of ego confrontation. Occasionally, however, spontaneous rational insights occur when the figures quickly interpret themselves through their own behavior, as in the following example of an 11-year-old girl:

> She imagined a male fox that was hunting ducks in the meadow. He was not successful. Instead, the female fox accompanying him soon caught several ducks. She then reproached him, "You could not succeed, because again you wanted to have all the ducks for yourself!" In that instant, the girl

opened her eyes and blurted out emotionally, "My brother always wants to have everything for himself, too." (29)

Preconscious and more emotionally colored insights are particularly effective when the image content represents resistances or frustrations, for instance, in the form of the already mentioned "obstructive motifs" or peculiar "absurdities" in the image content. Typical examples are situations in which the cow that is to be petted runs away; when the patient is confined within a barbed-wire fence and cannot find a gate; when the house has neither door nor windows; or when the brook is regularly dammed up. These kinds of confrontations with neurotic maladjustments, constrictions, or defensive processes often leave a certain degree of internal unrest behind, which is therapeutically effective. It goes without saying that the therapist is responsible for the dosage of such a confrontation. Subtle intuition and experience are necessary. We cannot elaborate further here on the details of these technical problems, which are relatively easy to master. Particular caution is required for each new case and for the beginner.

From application in practice, we drew the conclusion that even the search for diagnostically interesting standard motifs represents a part of therapy. Today, we recommend less frequently than previously a systematic psychodiagnostic procedural course and leave more therapeutic leeway for spontaneous development—for instance, in the training and free-associative procedures listed below. Therefore, it would be a misunderstanding to assume that each of the six standard motifs listed here must be used in the therapeutic technique. They are meant to serve only as points of orientation. It is more important to pay attention to the individual therapeutic tools, which are briefly described below.

THE THERAPEUTIC ASPECT

In presenting the *therapeutic aspect* of GAI, the special technique needs to be described and the stance of the therapist must be characterized. The theoretical foundations will be mentioned only in passing. We are assuming that the therapist is equipped with sufficient knowledge of the symbols of depth psychology and that he is well acquainted with the foundations of psychoanalysis, including the categories of transference and countertransference.

However, the present work cannot be regarded as a set of detailed instructions for conducting therapy. For these, one must refer to the original publications. It should suffice here to comment on the individual technical instruments and procedures. From the broad spectrum of techniques, the following are available for therapy with children:

1. Training procedure
2. Confrontation with symbols
3. The management model of symbolic drama using the techniques of "feeding and satiating" and "reconciliation"
4. Associative procedure

Training Procedure

The simplest technique, and also one that is accessible to the therapeutic trainee, is the *training procedure*. As previously described, the patient is led to the meadow and instructed to observe in detail. Through a series of questions he is urged to describe the surroundings, the weather, the presence of animals, and so forth. One can have him linger here, if it is a type of meadow worth exploring in more depth. The next step might be, for instance, a walk pursuing the course of a brook to its source. The patient can freshen up, get his face wet, have a drink of water, or perhaps go for a swim. In an additional training session one can encourage him to climb a mountain, to describe the panoramic view, and to return from there. In the next exercise one can attempt, for example, to have him follow the downstream course of a river and later to explore a house.

During these exercises the "obstructive motifs" previously described, which emerge as the expression of resistances and defensive processes, are registered attentively by the therapists. They are the actual dynamic aspect of the neurosis and, therefore, we welcome them as elements of the confrontation. We lead the patient along in carefully measured steps and have him repeatedly consider these motifs anew. As we cautiously explore, we can even offer solutions. In this process it is important never to use force. On the other hand, these motifs stimulate the patient to reflect, and they promote internal confrontation. The training procedure is relatively unspecific and does not require the exposure of the presenting problem. No matter what, the procedure occasionally brings about an unexpectedly rapid end to the symptoms, particularly those of anxiety neuroses and phobias, if phobically cathected situations are cautiously and realistically "practiced" in well-regulated steps. The following example illustrates the use of the training procedure:

An 11½-year-old student, Ilse, was sent to the child guidance center of a large city because of a striking deficiency in aptitude and poor school achievement. Although her IQ score was 126 (HAWIK, a Wechsler test for children in German), the patient's achievement was even below elementary-school level, and she had a pronounced test anxiety. A substantial handicap lay in the fact that the child was kept isolated from the neighborhood by her foreign-born mother and had learned the German language only a year before starting school. Extensive tests further showed that the central symptom was a persistent bridge phobia, which had existed for three years. Its psychodynamic history was not readily uncovered. However, the female therapist had the impression that—metaphorically speaking—a spanning of the "here" (representing the patient's own home) and the "there" (representing the German surroundings and the school atmosphere, perceived as hostile) was impossible and that therefore, the bridge phobia was a symbol of the patient's poor adjustment to the environment.

In therapy, autogenic training was taught first in the hope of at least diminishing the bridge phobia by developing intentional directives. However, even after four months of therapy, there was no improvement. The therapist then recalled GAI, to which she had been introduced through one of my courses. In the first session, she had Ilse visit the meadow. There, the girl soon found a small rivulet. The therapist had first attempted to have Ilse imagine a bridge. That attempt failed, however. In her imagination the patient was not able to cross even the small rivulet; she was anxiously inhibited. In the next session a shallow brook was visited, and the patient was asked to wade through it. She succeeded in doing that with no difficulty. In the third session a few larger stones were found in the same brook, with whose help the water could be more comfortably crossed, so that the patient could reach the other shore. In the fourth session Ilse was encouraged to find a footbridge over the same brook and to use it to get across. After some resistance she found the footbridge; however, it still lay in the water and had no railing. Thus, the child got her feet wet crossing over this footbridge. In the fifth and final session she managed to find a little bridge with a proper railing which brought the patient across the brook to the other shore.

Two days later the mother spontaneously called the therapist and stated with relief that the "terrible bridge anxiety" had finally disappeared. Suddenly, Ilse went across each of the numerous bridges in the city totally freely and naturally, as if nothing had ever been the matter. The therapist emphasized her initial skepticism with regard to a lasting success. She continued counseling the child because of the social difficulties mentioned. Ilse resisted all further "walks" in GAI, commenting that she was no longer afraid of bridges. Subsequently, the school problem continued to improve, and the child was freer in general and clearly more adjusted to her surroundings. The follow-up study a year later showed that the bridge anxiety had not reappeared during that period. The school situation had improved to the extent that Ilse was considered qualified to register for the entrance examination to the intermediate school. In the meantime, Ilse had surprisingly mastered many stresses that had arisen. In addition to school, at the age of 12 she managed the household independently while her mother lay sick in bed.*

The case is clinically remarkable in that an apparently very neurotically disturbed and socially constricted child could be freed in only five sessions from a strong bridge phobia that had lasted three years. In the training procedure, the therapist had gradually presented the resistance motif in steps regulated by the child, mediating and supporting her experience at each step. The child had been confronted with her anxiety and had been able to master it. The previous introduction to autogenic training and a strong positive transference to the therapist may have been favorable starting conditions.

A variant of the confrontation of anxiety-laden situations in steps can, of course, occur in an encounter with the patient's father and mother, preferably disguised as symbolic figures. With each symbolic figure that appears, whether it is an animal, a person, or a fairy-tale figure, the child can invent a conversation, ask questions and report the answers in order to initiate communication and clarification. In this way anxieties are reduced. A gradual approach to parental images that appear to be too powerful and dangerous can also be achieved in this way. We present an additional example:

The 10-year-old Joachim, a physician's son, was presented to a colleague working as a psychotherapist in a general practitioner's office. His presenting symptom was a persistent facial tic. In addition, the parents reported that the child was easily distracted by play and therefore did not come home after school and also could not concentrate in school. The somewhat shy and reserved child made the impression of being expressly well-behaved, anxious,

---

*I am indebted to Dr. Edda Klessmann, Lemgo, for the description of this case.

and inclined to daydream. Altogether, the demeanor of a small child was evident (27). Only during the GAI therapy did we learn more about the child-rearing regimen of the parents. The child was clearly afraid of the parents, who were bringing him up in a very strict and restrictive manner. Drastic measures were taken after only minor misbehaviors. For example, the child was given several days of house arrest after coming home only a half hour too late.

In GAI numerous scenes emerged with animals in the meadow: an elephant, a lion, an octopus, a dragon, and a cow as variants of symbolic representatives of the powerful parental figures. At first the child fled from them. In his imagination he turned to smaller animals, such as cats, mice, hares, young foxes, badgers, and beetles and played with them. Later, the therapist attempted a confrontation with the more powerful animals. This confrontation was continued using the techniques of feeding and satiating (cf. below). Joachim's fear of the strict parental figures seemed to diminish gradually. On their occasional visits during this period, the parents were informed of the childlike contents of Joachim's affective imagery. They were urged to relax the pressure of their upbringing methods. Impressed by the evidence provided by the childlike imaginings, they readily accepted this advice. People who visited the parental home of the small patient soon reported with astonishment that the "table manners" had been "relaxed."

As treatment continued, the house was explored. Marionette dolls of Caspar, King and Witch, whose strings had got entangled with each other, were found in a trunk in the cellar. The relationship between the parents and the child seemed to be expressed by this entanglement. Then the therapist encouraged the patient to carefully disentangle the strings of the marionettes.

Toward the end of the treatment, fighting scenes finally developed, in which the possibilities of aggressive assertion were realized: Smaller animals fought larger ones and finally conquered them. Death wishes were also symbolized in this way. At the end of treatment, a shabbily dressed royal couple appeared, who gave the patient gifts. In this image, the therapist perceived the patient's wish to dethrone the parents or to decrease their power and, at the same time, to get attention and strength from them.

In the course of treatment a repeatedly appearing forest was used as a control of the synchronous change. At the beginning the trees generally had neither crowns nor leaves. Step by step, the image became normalized, and at the end, the trees had full crowns and leaves.

The therapy lasted from July to September 1963. In the total of three months, 11 sessions were used for GAI and 6 for introductory discussions and other dialoging therapy. The follow-up study three years later showed that the tic had never appeared again. The child also seemed to have changed in his total behavior. He was less anxious, more fresh, and more boyish and communicative.*

---

*I am indebted to Dr. Maria Beck, Darmstadt, for the description of this case.

In this case, the treatment was conducted within the framework of the technique of symbolic drama (see below), whereby very definite goals were set, namely, the progressive confrontation with the overwhelming parental images by means of techniques of confrontation, of feeding and satiating and of reconciliation to the point of an assertive exchange. The result was the disempowering of the parental figures, who were invested with infantile misinterpretation and projections. The therapist had proceeded judiciously and had also not been misled by the disguise as animal and fairy-tale figures (king and queen) into slipping into this infantile world but kept in mind the goal of improvement through realistic adjustment.

The double-track nature of the therapeutic process in child guidance is particularly obvious in this case. One is often surprised by how relatively easy it is to influence child-rearing principles of cooperative parents, when one demonstrates and explains to them the child's fantasies that concern their methods of child rearing. This potential influence, provided by the vividness of the catathymic imagery, seems to be one of the advantages of the procedure in the therapy of children and adolescents.

Confrontation with Symbols

With this therapeutic procedure, we aim at situations in which the child is cautiously brought face to face with anxiety-producing symbolic figures and is induced to confront them. In the preceding case, this confrontation occurred with the parental images. The therapist first asks the patient to observe the threatening figure, whether it appears in the meadow or from the woods—if need be, with the patient hiding behind a bush. The figure should be described in detail, particularly the facial features and the expression of the eyes (especially important for children with hysterical structure). Thus, the anxiety affect is disclosed. This disclosure can be tolerated under the therapist's protection and is mastered by an analysis of the image, which leads to a more mature form of ego control. The therapist can help by encouraging the child or by energetically urging him to look directly—not to run away—and to conquer his anxiety. The therapist can also assure the child that he is standing directly behind him (in GAI), thus offering him protection at all times. The therapist can even actually hold the child's hand under certain circumstances. In this way, the child can courageously face hostile figures. Then, he should banish the figure with his own look, an almost magical act, for which he is given instructions. The simplest way of making the situation understandable to the child is to refer to the Grimms' fairy tale entitled "The Youth Who Could Not Shudder." This method of confrontation has, perhaps, a certain heroic accent; it requires ego strength on the part of the patient and sufficient experience on the part of the therapist. The technique of confronting symbols can be advantageously supplemented by directions to feed and satiate in the symbolic drama. In this way, the child can pacify a hostile symbolic figure by feeding it. For example, a saucer of milk is placed before a snake, honey before a bear; one throws hay to a cow and invites a person to breakfast.

The Management Model of Symbolic Drama

In treating children and adolescents, the technique of feeding and satiating just suggested is the mildest way of approaching symbolic figures that arouse anxiety. It is important that they be fed to excess; they then become tired and powerless and finally act in a more friendly manner so that a reconciliation can be initiated later:

A 6-year-old boy awoke terrified because a giant threatened him in a dream. In GAI, the child was advised to transport truckloads of cattle and to offer them to the giant. After some hesitation, the giant grabbed the cattle and was encouraged to eat more and more. Finally his belly was full, and he became fat and tired and lay down to sleep. As a result the child went to bed having abandoned this kind of dream.

This example illustrates the following application:

1. The revival of a night dream in GAI as an additional directive technique. Although its meaning is limited, often, a significant dream can be used as the starting point for a GAI session, to bring a pressing problem to imaginal representation and thus to a working through. In a relaxed state, the child is asked to recall the last image of the dream and to develop it further. In this way, the therapist can steer the dream in a therapeutic direction, unless he prefers to let the dream "unfold" by itself freely.

2. An additional directive model is reconciliation and tender embracing, in which the patient comes into physical contact with the

symbolic figure, as in stroking, which makes a genuine reconciliation possible. This procedure is related to feeding and satiating and can lead to a conclusive assimilation of hostile, split-off contents of a complex. Of course, anxieties can also be liberated in this way.

The effect of all of these forms of symbolic drama is often surprising, sometimes dramatic. One example, which comes from a student in the first semester of his studies, may serve as an illustration.

A 21-year-old chemistry student was unable to complete an examination given by his professor, although he had previously done well on a test given by a lower-ranking faculty member. From his previous history, it was clear that the patient had projected the image of his domineering father, whom he perceived as outstanding, onto the professor. We also learned that the patient had also seen his former chemistry teacher as a father figure: like a strict high school principal who was admired with anxious reserve.

Using symbolic drama, we asked the patient to imagine this elderly gentleman stepping out of the darkness of the woods into the meadow. After long hesitation and much encouragement (resistance), the former chemistry teacher finally appeared with a grouchy and rejecting look on his face. Taking no notice of the patient, he went on his way. We asked the patient to observe him precisely (confrontation). Finally, we suggested that the patient begin an imaginary conversation with the former teacher. At first, the patient accompanied the teacher reluctantly, experiencing some anxiety. However, after contact had been made, we asked the patient to unpack his picnic lunch (we had already suggested that he fill his pockets with good things to eat). At first, the former teacher refused indignantly (resistance). Finally, however, we succeeded in getting the two to sit down and eat together. Soon the teacher was eagerly eating the offered food, and together, student and teacher drank a bottle of red wine. After that, the teacher became much more friendly and finally they slapped each other on the shoulder (first, feeding and satiating; then reconciliation and tender embrace).

At the end of this three quarter-hour session, the patient was advised to again imagine, before going to bed, that he was giving the teacher something to eat and was becoming reconciled with him. The excessive amount of food was important. The next day, the young student passed the examination with his professor without any difficulty.

With regard to psychodynamic perspectives, it can be briefly pointed out that with this procedure a substitute father figure, who had previously been defended against and isolated, was now reintegrated and apparently assimilated by the release of anxieties. However at the same time the chemistry teacher, who was connected with the image of the father, was separated from the profoundly relevant infantile father image. In correspondence with our clinical experience using the technique described here, it is possible to release infantile projections and make possible contrasts with the infantile base without working through all of the infantile parental images. The process of thoroughly resolving infantile parental images requires substantially longer treatment periods in GAI as with other therapies.

We will not present further models of symbolic drama here. The most important models for this age group are "feeding and satiating" and "reconciliation and tender embracing." Third in importance is symbolic confrontation. In view of the difficulty in estimating the appropriate intensity of confrontation, symbolic confrontation should only be used by an experienced therapist.

The Associative Procedure

This procedure, which was the last one we developed (17), replaces free association in the Freudian sense. Instead of associating thoughts, we have chains of freely developing images. For example, after the meadow motif has been introduced, further image developments follow spontaneously, unguided by the therapist. These image associations occur most easily in somewhat hysterical people and also in children. They create further scenic images out of the joy of making up stories. The images tend to circle the central problem in constantly changing versions and to converge there. Initially, explicit, symbolic disguises make it impossible for the patient to know consciously which problem is presenting itself to him. When he confronts the father problem, for instance, images such as dark clouds or high mountains may appear at the beginning. As therapy progresses, the motifs gradually change from inanimate to animate objects. The associative procedure can be combined with the other technical procedures already mentioned, as soon as the patient spontaneously produces a chain of images, so that interventions by the therapist are avoided. Of course, resistances are sometimes skillfully circumnavigated by this means, and new kinds of reactions are constructed, for instance, to please the therapist. Then, more directive measures must be instituted.

The Behavior of the Therapist

In conclusion, we would like to discuss the therapist's behavior. Ordinarily, as in play therapy, the therapist's stance should be consistent and should be based on a reserved and giving friendliness. Otherwise, the therapist should remain predominantly passive, awaiting the patient's spontaneous productions. According to the type of case, the therapist cautiously intervenes in and guides the affective imagery by suggesting certain motifs and recommending certain tasks. At the start of treatment, he will, above all, instruct the patient to describe in detail all that the patient experiences. In this way, the patient's current problems can be focused on through the choice of appropriate motifs, as in the example of the child with the bridge phobia, who crossed the brook. The therapist remains at all times fully aware that all of the imaginal motifs have a *symbolic character*. Accordingly, everything that he does to direct the patient is also symbolic and is a direct intervention into preconscious emotional constellations (see "Operation on the Symbol" 13). One must also recognize that the feelings accompanying or set free by the images are as therapeutically significant as the production of images itself. However, emotions and affects manifest themselves in a sluggish flow. Similarly, GAI also frequently occurs sluggishly or haltingly and may be interrupted by pauses. These pauses are often significant and indicate that affective impulses are rising or subsiding. The patient is then strongly emotionally involved, even if in the form of a defense. With an older child it is advantageous to allow the feelings that arise to become conscious by urging him to describe what he is feeling. We also encourage the patient to describe the emotional tone of the meadow or landscape motif, for instance, of the view from a mountain or the expression of a face, and if necessary, we ask about analogous feelings or memories. Young people in puberty, especially, who defend against their emotional lives very strongly, can be encouraged to perceive and express feelings in this way.

Questions about interpreting the contents of GAI are repeatedly asked. Interpretations were almost never attempted in the cases described here. In play therapy, one proceeds on the assumption that the scenic developments and the contents themselves directly demonstrate the problem to the patient over a long period of time, even if only on a preconscious level. In this respect, GAI is also a self-interpreting procedure. In our earlier publications, GAI was called a *nonanalyzing* or *noninterpreting* procedure. Particularly for children and adolescents, the noninterpreting process is a legitimate one. An experienced therapist will be able to *offer* an interpretation at important points that signal a typical behavioral disorder. Otherwise, the detailed descriptions of the catathymic contents, which the therapist repeatedly encourages, and the verbalizations of the corresponding feeling tones are the instruments of clarification that raise the experiential material to the threshold of consciousness. The rest is left to spontaneous inspiration. Only in long-term treatment is a broader analysis, bringing more into consciousness, necessary, along with an analysis of the transference resistances and of neurotic character resistances. Our procedure gains a somewhat more active accent through the technique of symbolic confrontation, for instance, when the child is face to face with an anxiety-arousing animal and is held fast, encouraged, and at the same time protected.

In general, however, we tend more and more toward stengthening the mature parts of the ego and not toward encouraging the patient's dependency, for instance, by the offer of engulfing protection and "love." These few words should suffice.

Within the framework of this book, we cannot go into detail with regard to the theoretical positions of GAI or to the obvious objections from the point of view of classical psychoanalytic technique. Such a discussion will appear in the overall total presentation of the procedure that is now being planned.

Summary

In summarizing, we find that GAI, as an extensively studied method of employing mental imagery in psychotherapy, is particularly suited for the treatment of children and adolescents. Whether one emphasizes the projective character of the hypnagogic visions that are released or the free productions of fantasy, we are dealing with a psychoanalytically oriented method. The parallel with play therapy is evident. For children, the main indications for GAI lie in those age groups that are no longer open to play therapy and are not yet amenable to adult analysis. For practical purposes, however, the procedure can be used from age 7 up to adulthood.

The technique was presented in detail and illustrated with the aid of examples. Individual case results were demonstrated in order to show that GAI can serve as short-term therapy. Symptoms that have existed for years can be eliminated, and a change in personality can be initiated. The results held up in follow-up studies carried out up to three years after therapy. The parents can be influenced to abandon faulty child-rearing principles, if their child's imagery is explained to them and interpreted for them.

3

Special Features of Working with Guided Affective Imagery in the Treatment of Children and Adolescents

EDDA KLESSMANN

General Technique

This introduction concentrates on only a few aspects in order to avoid repetitions. Guided affective imagery (GAI) can be made accessible to therapists of the most diverse origin and it can be successfully applied by them after adequate training. As always, however, basic knowledge and experience in psychodynamics are essential, for instance, for comprehending symbolic language and for guiding the patient, taking into consideration the transference-countertransference relationships.

Combinations of GAI, which often facilitates profound therapeutic experiences, with complimentary or similar procedures are possible and productive. So it is that GAI can be employed by therapists who function within the perspectives of behavior therapy. Further, in GAI, one can "accompany" the patient as one does in conversationally-oriented person-centered (client-centered) therapy. Elements from the psychodrama configuration can also be incorporated into GAI, or, conversely, imagined scenes can be represented in the form of symboldrama as Wächter (58) has shown.

In connection with the psychodrama, I will cite a personal statement made by Leuner that he successfully used role playing within GAI in treating a female teacher. The patient had the habit of calling on her inattentive students with pointed questions. After the role play, in which the patient had taken the role of a girl student, she became conscious of the unpleasantly harassing quality of her method.

Finally, there is the possibility of "satisfying" the backlog demands of early suppressed-drive needs. (Two examples from the the therapy of a case of anorexia nervosa will be given later.) A further indication of the extent of variation possible with GAI lies in its use with children and adolescents in groups, both with and without music. GAI offers the possibilities of modification necessary for effective treatment appropriate both to the therapeutic situation and to the structure or accessibility of the patient, as well as to the level of training and the structure of the therapist. Not infrequently, astonishingly fast and lasting improvements, above all in behavioral disorders of children, can be attained with GAI and its flexible application.

Special Features of GAI with Children and Adolescents

Generally, the heightened suggestibility of young patients is among the special features of the use of GAI with children and adolescents. Thus, as a rule, children can be easily guided in GAI, but I would like especially to emphasize that they can also be misled.* For this reason, a good and trusting basis for transference must be present. The therapist must know to what extent anxieties can be mobilized and worked through in order to avoid undesirable incidents.

A child's neurotic structure is distinguished by the fact that the defense mechanisms are usually less rigid. Compulsive symptoms as well as anxiety ideations, for example, can often be resolved amazingly quickly. Here, symbolic drama is directly indicated. In my experience, the child's symbolism is, in general, structured more simply and is easier to understand than that of the adult. Nevertheless, one must expect that even with younger children, more complex affective images—for instance, archetypal contents—can arise.

With children as well as with adolescents, a rapid change of scene and of transference frequently occurs. Thus, it is often impossible to proceed exclusively with training, symbolic drama, or associa-

---

*A 9-year-old patient of mine, for instance, lost her way in a cave that was getting darker and darker, until she suddenly opened her eyes, both startled and confused. (At that time, I had had very little experience with GAI.)

tions. It can also happen that "magical fluids," such as saliva, are used spontaneously and very early in treatment to overcome anxiety. Commensurate flexibility on the therapist's part is therefore necessary. On the other hand, the therapist must be responsible for at least a minimum of continuity and calm guidance. In the course of therapy, the child must also be progressively confronted with his anxieties. However, he can confront these anxieties only if he never feels left alone with his GAI experiences. Leuner (see Chapter 2) defined the therapist's protective function as follows: "To strengthen the mature parts of the ego," but "not to promote the patient's dependency by the overflowing offer of protection and 'love.' "

Whereas Leuner advised adults to write down their GAI impressions after the sessions in order to reinforce them, I recommend that children draw their affective images.

In the last two years, I have increasingly changed to treating younger schoolchildren (approximately 6- to 9-year-olds) in a sitting position. Apparently, this position is more appropriate to this motorically active age-group. With their eyes closed, children of this age occasionally make involuntary movements with their arms or legs. Probably, these movements also reduce the anxieties about being at the therapist's mercy, which occasionally arise in the prone position.

Indications and Contraindications

Childhood phobias can be influenced especially well by GAI. With phobias, gradual "deconditioning," approximately according to the principle of behavior therapy, proves successful (see in Chapter 2, treatment of a bridge phobia, p. 31 ff.). There have been occasional successes with compulsions in "retrieving" the material that was split off from consciousness (i.e., repressed), bringing it into the affective image, and making the symptom comprehensible through a corresponding interpretation. In this way, one can put an end to a "meaningless" repetition compulsion, for instance in the ritual of an attempted reconciliation, by working through the underlying guilt feelings.

A child, who in GAI could scarcely resist the compulsive urge to have to pick up a handkerchief lying on the rug (the handkerchief seemed to be "alive" to her), learned to understand that she had to "make up for" an early omission toward her younger brother over and over again. From that point of on, she was able to ignore the handkerchief. She drew the test in Figure 4.

Fundamentally, GAI can be applied in all cases of children's behavioral abnormalities, beginning, for instance, with enuresis and persistent stuttering on up to severe psychosomatic illnesses such as anorexia nervosa (see p. 79 ff.).

In my opinion, there are no clear-cut contraindications within the framework of emotionally conditioned behavioral disturbances. Caution is advised with rapidly developing anxiety states and severe depressive moods.

Naturally, there are cases that cannot be reached by GAI. Early childhood pathological habits (such as jactitation) serve as a limit, as do the more severe levels of feeblemindedness or organic brain syndrome. The procedure is not to be used with childhood psychoses.

It seems important to me to note that children's therapies occasionally stagnate in general, because pathogenic family influences are not eliminated. A delayed GAI "starter" may document this:

The 13-year-old Rolf, a later child of relatively old parents, was presented at the guidance center, because for three months following the death of his (maternal) grandfather, he would not go anywhere near the cemetery, not to mention enter the cemetery itself. During the same period, he also would not sleep in his room. After the death of his (maternal) grandmother two months later, he made a brief attempt to sleep in his room again, then went crying to what had been established in the meantime, as his spot in his parents' bed and explained that he could not sleep alone anymore. He was afraid that his (deceased) grandmother would come into his room.

The mother herself had not yet got over the death of her parents. Each time the subject came up, her eyes filled with tears.

Besides this anxiety ideation, a pronounced "aquaphobia," existing over a longer period of time, was reported. After the 4-year-old daughter of a cousin of the mother had drowned, the boy's parents had signed him up for swimming lessons with the reasoning, "You see, that's why one has to learn to swim!" Nevertheless (!), Rolf always trembled with fear and still cannot swim to this day. He claimed he could not get any air in the water.

Accordingly, the diagnosis reads: Numerous traumatically caused phobias, faulty identification in sex role (see below). It also needs to be mentioned from his previous history that the boy has always been very excitable.

Occasionally, he has vomited in the morning before school. After his grandparents' deaths, he became "so crazy" and worsened in school to the extent that consideration was being given to transferring him to a lower-level school in spite of his talent.

The weak, adipose boy, making a girlish impression, seemed to be intensely fixated on his mother. She regularly brought him to his sessions, although he was fully capable of coming on his own. We learned from her that he had a grade of "B" in the subject "needlework." He enjoyed embroidering doilies very much and was very exact in doing so. In general, he was rather "fussy" and particular about his things (Figure 1).

In view of the relatively short duration of the current symptoms (cemetery fear) GAI was initiated after autogenic training, which had brought only temporary successes in school. A short-term therapy of cautiously progressive confrontations was planned.

Signs of constriction (anxiety) were present in the "meadow" in the form of barbed-wire fences. It was interesting that in the attempt to wade through a brook, Rolf was clearly afraid of not being able to get any air (swimming lessons). However, in a later attempt, he was able to overcome this fear (fourth GAI session). A "romantic house" (see Figure 1) could symbolize his clearly faulty feminine identification. During a trip to Helgoland, which he chose himself in the fourth GAI session (a week earlier, he had actually taken such a trip with his parents), his aquaphobia became evident. However, for the first time, he succeeded in identifying with his father in GAI. He went on

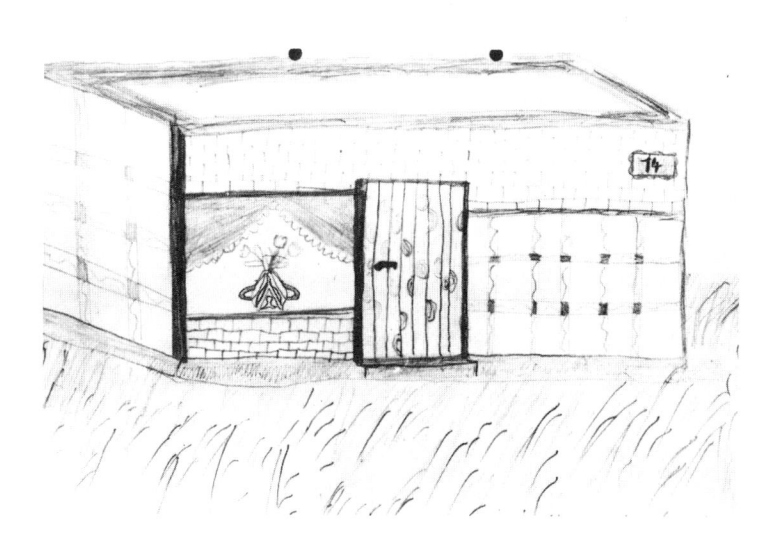

FIGURE 1: "Romantic house" of a 13-year-old boy with feminine identification.

expeditions with him onboard a ship, bought peanut chips and fed the gulls. He eagerly took photographs, while his mother "chose to go inside" (in reality, on the Helgoland trip, the mother had sat down on the boy's lap and had clung to Rolf when it turned stormy). In the fifth GAI session, he chose the theme "school" and after some initial irritation was able to overcome his anxieties and "craziness" about an assignment.

Finally, in the sixth GAI session, I suggested the cemetery theme. Rolf spontaneously chose to accompany a friend (a "pacemaker" as a management model in symbolic drama according to Leuner). At first, I perceived clear resistances, pauses, and uncertainties. Finally, he stated, "I want to go back now." However, he then followed his friend after all, who was leading the way, read the name of someone he did not know on a gravestone and finally discovered that he could not read the name on "grandma's grave." "It is somewhat covered over with blue flowers." After this confrontation, he was visibly freer and promised that he would actually visit the cemetery with his friend within the next few days.

To my surprise, he felt distinctly worse in the next hour. He refused to go to the cemetery in GAI. He had also become more restless again and, as I later learned, had gone back to sleeping in his parents' bed, which he had previously given up. In a conversation with his mother, I learned that she had been very shocked and disappointed when, after the last session, Rolf joyfully told her that he now wanted to go to the cemetery with a friend. It was the first anniversary of the grandfather's death, and the mother was obviously depressed. Her disappointment was expressed in the words, "Why don't you go to the cemetery with me?" In this request, she was unconsciously expecting, as on the Helgoland trip, care and protection from her son, as the father always avoided such "sentimental requests." I now also learned that when her son practiced autogenic training in the evening, she sat next to him with a stopwatch, "so that he would practice well."

Subsequently the cemetery theme and the family sleeping arrangements were discussed with the mother. In the meantime, she herself had noticed that the child's symptoms were closely connected with her own problems and had, partly on my advice, begun a course in autogenic training.

When I saw Rolf and his mother for a consultation two months later, the mother was substantially freer. The son reported that he had not only successfully completed his swimming course, but had also been to the cemetery five times with his friend without being afraid, "as here with you in the daydream." After that, I dispensed with a continuation of the therapy. However, I came to the striking realization that neglecting the immediate family realities can block even the most perfect GAI therapy. Only the working through of the obstructive surroundings produced the desired therapeutic success. Still, with the relatively acute symptoms, this success was realized after only six GAI sessions, one per week.

Additional Case Examples

Cases of Children

I would like to begin the next case presentation with the description of a drawing of a tree by a 12-year-old girl who was suffering from anorexia nervosa. She drew a round-topped tree with free-floating fruit; they were cherries. The tree had no branch or leaf structure whatsoever in the crown. The drawing had been completed one year before GAI. At that time, I had taken over the child's aftercare treatment from a clinic. In the scenotest, the child presented a family structure typical of anorectic patients. Therefore, the therapy began with a role play taken from scenomaterial (scenodrama). The child had worked through the family and role conflict well in this material. She had also become much freer otherwise. At the termination of treatment, there were no longer any signs of anorexia nervosa present. For that reason, I wanted to perform a "trial GAI" as a control for the one-year follow-up examination. The child spontaneously pictured a round-topped apple tree in the meadow. At closer observation, the tree corresponded surprisingly with the test tree drawn one year previously. Instead of the free-floating cherries, red apples could now be seen in the GAI. In addition to the corresponding fruit trees, the fact that the patient experienced a strikingly large number of obstructive motifs was also interesting, whereas she otherwise seemed to be internally free and largely symptom-free. In my opinion, it could therefore be assumed that in the GAI experimental layer, deeper-lying disorders are registered. In a certain sense, the GAI had a control function. Contrary to my original intention I kept the child under further observation.

I treated the next case exclusively with GAI. Aside from practice in autogenic training, no additional techniques were applied. This therapy, which required 18 sessions, can in my opinion be considered a typical example of "children's GAI":

Hannes was a 9-year-old boy suffering from dyslexia. His mother had passed away when he was 5 years old. During the mother's 3-year illness and also after her death, Hannes had been influenced by various substitute mothers, who mainly spoiled him with material things. Between two dominant sisters, he could only assert himself with difficulty and tried to circumvent any arising problems with his soft, submissive demeanor. To compensate for his constant school disappointments, he found a substitute gratification: Instead of going to school, he bought chocolate with stolen money and went for walks.

The first GAI session vividly reproduced the child's anxious-depressive condition. Hannes imagined a rectangular barren meadow surrounded by barbed wire. In the middle stood a pole approximately one-half meter high. To the left of the meadow, there was an uncanny, dark forest and a village in the distance. It was impossible for him to imagine a brook or a mountain. Therefore, I finally had him enter the village. There, he discovered only empty streets, no people, no animals or trees (according to Leuner, 38 , this emptiness should be assessed as a depressive sign). But there was a church there. In the churchtower, there were two windows; in the "house"—he meant the nave—were three large windows. (Symbolically, one could assume that these objects were the parents; the father, who had remarried in the meantime, elevated in the tower windows and the three siblings "one floor below.")

Attempting to look in cautiously through the church door, Hannes perceived an altar and behind it a large picture, partly colored, partly black, as well as empty benches.

Later, Hannes walked down the lonely village street and came to a rectangular lake, which was completely surrounded by impenetrable woods. When he plunged a barbed-wire post into it, he found out that the lake was "awfully deep."

In an adult, such a desolate catathymic landscape signaling depression would disquiet the therapist. With children, however, one can expect amazing transformations. Therefore, I had Hannes continue to fantasize in the following session. This session also brought little that was positive. Instead of a meadow, he spontaneously imagined himself in front of dark woods, which stood before him like a wall. There was an endlessly wide field next to it and nothing else. In such cases, I try to have the patient focus on a particular part of the image. By keeping detailed eye contact with the world, one often succeeds in overcoming an unconnectedness with the environment. Therefore, I asked Hannes whether he could perhaps discover a small animal or something similar, by looking at a particular tree more closely. Finally, he did see an animal which turned out to be an ant. Eye contact with it was, of course, not possible. Thus, this session also did not turn out to be very encouraging.

When the third session began, it was influenced by a different event. In school, Hannes had received one of his usual "Ds" or "Fs" again. His imagination in this session, in its childlike and concrete simplicity, stood in contrast to the pretentious introductory theme of the church symbolism. On the left-hand side of a blackboard (Figure 2) 4s ascend in the form of an arcade, and 5s and 6s appear in a downward curve. (In the German school system, grades are given in numbers rather than letters, "1" being the highest.) A red mark, the correction made by the teacher, connected the deplorable series of numbers. In GAI, the unattainable grades appeared above the board in gleaming white in the sequences 3, 2, 1—rising diagonally upward. The animal beneath his image of grades was a dog, the first living being that had appeared in GAI by Hannes. Since infancy, he had suffered from a pro-

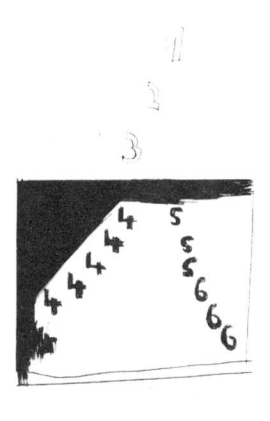

FIGURE 2: First appearance of the "anxiety objects" (poor grades and a dog) in the GAI of a 9-year-old dyslexic patient. (The blackboard was not colored totally black; because of a lack of time, the session was ended before this could be done.)

nounced dog phobia. Through the emergence of the feared object, for the first time, the possibility arose of proceeding with some symbolic drama in order to achieve an increasing strengthening of the ego by confronting the object cathected with anxiety. This procedure succeeded and, in reality, also had the effect of diminishing the dog phobia. In symbolic drama, Hannes now began to approach the figures of the parental images, whom he symbolically imagined as a cow and a bull. From the representation, which he subsequently drew, it became clear how he gradually conquered the terrain. Separated by barbed wire, he initially stood outside of the scenery as a "small point." In GAI, through the principle of feeding, he first attained a reconciliation and confidence in dealing with the animals; the cow and the bull had been amply fed. This process finally allowed him to get into such high spirits that he rode on the cow's back and by pulling its right ear caused it to turn in circles, as he wished. The confrontation with the bull took place in an arena (Figure 3a/b).

After terminating therapy, Hannes made positive steps both in general and in school. He later attended and successfully graduated from high school and showed no further neurotic symptoms. Ten years have passed in the meantime.

In conclusion, I would like to discuss a modification of GAI, that has recently proved successful for me in working through uncon-

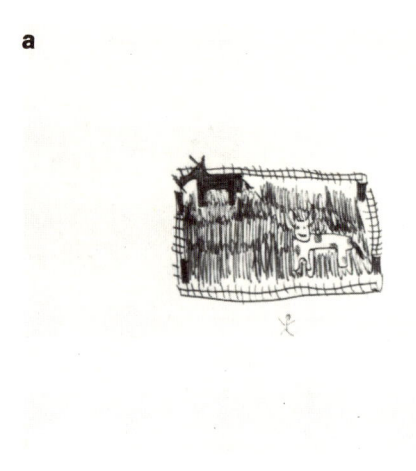

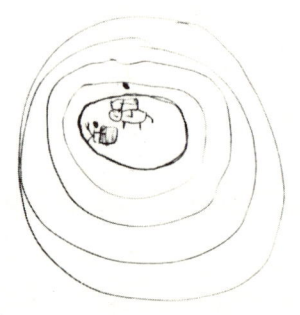

FIGURE 3a and 3b: Transformation through symbolic confrontation in the same patient. Initially, he stood helplessly in front of a barbed-wire fence facing "mother–cow" and "father–bull." At the termination of therapy, an open confrontation with the bull took place in the arena.

quered family conflicts with children. For a full year, I have been giving a "three-tree-test," which frequently proves to be a good projective test for intrafamilial conflictual constellations. The child is instructed to draw any three trees on a piece of paper and then to compare these to the people with whom he is most closely related, his family (Figure 4). If the representations are full of conflict, I generally bring the test picture into the GAI again and have the children make contact with the trees. This procedure often succeeds better than attempts with the imagined parents or with symbolic animal-parental figures. The child usually grasps his own role in this dynamic field without comment and is able to find possibilities for solving and mastering problems in the desired symbolic confrontation.

The 12-year-old girl who drew the three trees shown in Figure 4, tried to climb the apple tree (younger brother) and to get an apple for herself. However, the tree explained to her that this hurt him. Then, she slid down and asked the "pine-father-tree" for advice. Among other things she learned from him that he and the fir tree (she herself!) had "always" stood there, whereas the apple tree had come later. The patient followed the pine tree's advice and asked the apple tree for an apple. Now, she received what she wanted voluntarily. Subsequently, she also found better possibilities in reality for getting along with her brother-rival. Here, we see a kind of "GAI family therapy."

Cases of Adolescents

Following this small collection of examples of children's cases, I would like to go on to a few features that are peculiar to adolescents.

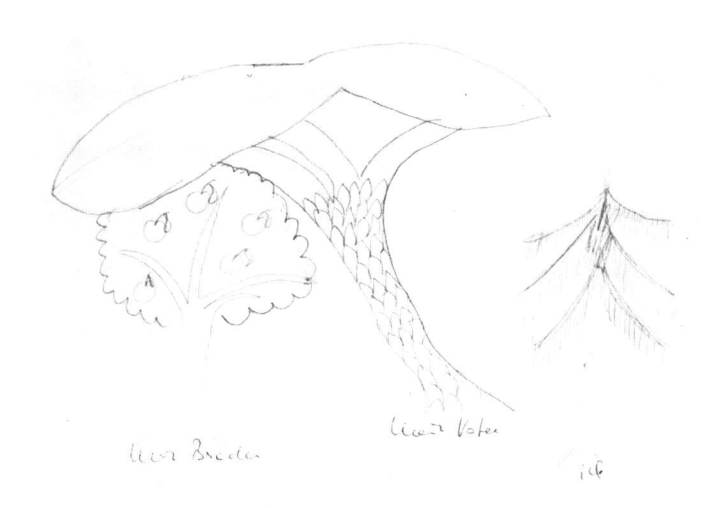

FIGURE 4: "Three-tree-test" of a 12-year-old obsessive-compulsive, neurotic girl, which was "worked through" later in GAI (see text).

With this age group, one must count on atypical developments in therapy more frequently.

In the first case, I will trace the spontaneous crisis intervention of a 17-year-old high school student, Claudia. A dramatic session was prefaced by an initially seemingly harmless meadow. I already knew Claudia from a forensic report (manifest incest with her father). While the father was serving his sentence in a penitentiary, Claudia developed a deep guilt complex. She had not consciously brought about the father's sexual advance, but because she had passively allowed it, she did not feel innocent either. During the three days before she came to our guidance center, she had not got out of bed because of an extreme state of anxiety and restlessness. I conducted a GAI. At first, she imagined a meadow that was familiar to her, where she had once been with her boyfriend. The meadow was bordered on the right by a brook. There was a forest in the background. As the patient tried to cross a river in a small boat, an insurmountable dam suddenly appeared in front of her. A fence, situated opposite her, which at first seemed to be harmless, turned out to be electrically charged. The forest in the background became threateningly thick. Suddenly, a house in front of it no longer had windows or doors, and it blocked the final way out. Claudia felt as if she were in prison. She reacted with distinct excitatory vegetative signs: intense breathing and red spots on her neck. I reminded Claudia that she had mentioned her boyfriend beforehand. She then produced a chain of associations recollecting material that she had previously repressed. She talked about sexual advances made by her boyfriend, which she had experienced very discrepantly. She remembered events from her childhood, which she had never mentioned before. After this strongly emotion laden report, she became visibly calmer (abreaction after age regression). When she was then urged to look around the imagined meadow again, it looked "harmless" as in the first image. The phobic state, which had cause her to enter treatment, subsided after this session.

With another 17-year-old female patient, the motif of the meadow also proved to be significant and led to the solution of an intensely reactivated, apparently central conflict. Brigitte was in the midst of a seemingly relatively harmless maturational crisis. We called the treatment with her the "struggle of the meadows." As in a double-exposure film, as she herself labeled the situation, two quite different meadows appeared. One represented an actual meadow from her childhood, where Brigitte had often played. The other was a product of her fantasy, her "future meadow." It was covered with daisies and was where she wanted to go for a walk with her boyfriend. The two meadows constantly alternated with each other. This alteration exactly corresponded with the patient's conflict: on the one hand, continuing to stay under parental protection, and on the other hand, separating from her parents and searching for her own, still uncertain, path with her boyfriend. Since, in GAI, she was totally fixated on this struggle for dominance between the meadows, I suggested that she again concentrate on the two meadows at home in order to gradually realize in which of the meadows she felt better. She then finally decided on the "future meadow" and seemed to have mastered a substantial part of her crisis of separation from her parental home.

In this case, I was able to solve the immediate problem, which had spontaneously crystallized in the meadow motif, that is, directly after the simple training program at the elementary level. Naturally, that is not regularly the case. Also, only in rare cases do I have patients "practice" comprehensive motifs at home alone, such as the meadows. In addition, I would like to urgently caution against having children and adolescents take "solo runs" at home, and as a rule, I inform my patients accordingly.

In the beginning of this chapter I indicated that emotional "backlog needs" can be satisfied in GAI.

Heidi, the 15-year-old, suffered from anorexia nervosa. After intensively working through the oral theme, she apparently had to actualize a part of the strongly repressed area of anal drives and reexperience it. She had been brought up to be extremely clean, and up to that time, her handwriting style had been strikingly sharp, clear, and seemingly compulsive.

The first breakthrough of her repressed anal impulses in GAI occurred when she unexpectedly formed a lump out of mud and threw it against a glass partition. This formed a strange pattern, which she subsequently sketched. She had thrown additional lumps of mud against the partition, alternating "two above and two below." These reminded her of breasts and the labia of the vulva. Between these, she placed three mudballs vertically, one above the other, and called the formation "stem." The connection with the male genitals did not occur to her. They were still strictly taboo for her at that time. Instead, she commented on the overall picture, which had a somewhat arched form, calling it "shit breasts." With this half laughing, half angrily devaluing outcry, she began for the first time to confront the "female"—meaning maternal and thereby her own—"inferiority." After this single anal breakthrough in GAI, her own cramped good behavior disappeared. Her handwriting became substantially more relaxed and larger. A letter written at this point documented her subsequent quite unaffected style as well as her enjoyment in continuing to indulge in anal terminology. *Shitty* was repeated three times as a heading applied to the "summing up of a week that was full of disappointments," the subtitle of her letter. After that, the patient's chronic obstipation also improved.

Through the case of the 18-year-old high school student Christoph, whom I treated, I would like to depict the unburdening function of GAI, repeatedly emphasized by Leuner, even in more-or-less clearly recognizable borderline psychotic cases. He had been so tormented by continual fantasies, dealing predominantly with the castration theme, that he was afraid for some time that he was losing his mind or becoming schizophrenic. He had physical sensations as well and believed, among other things, that his head would burst. He walked through the city parks for hours, even at night. In compulsive-phobic form, he avoided the consumption of alcohol and cigarettes, which he at times used again as a stimulant in an almost addictive fashion. Subsequently, he regularly became panicky, afraid that his "ganglia cells would be destroyed." "Crazy anxiety states" also appeared cyclically, which led him to fear that he was losing his eyesight. At these times, he felt "very strange sensations" and pains in his eyes, which seemed always to confirm that his fears of becoming blind were justified. In a compulsive manner, he had developed a magic counting system to defend against these anxieties. At night, he moved his bed to the center of the room, because he did not want to sleep next to the wall. He was afraid that the wall would fall on his head and damage his brain. Behind the fear of losing his intellectual potency, fears about his masculine potency, in general, increasingly shone through. Becoming blind became a major symptom of his apprehensions.

For a long period during therapy, he experienced both in dreams and in GAI that his eyeballs were being pierced with sharp instruments. The patient was familiar with this motif from the Oedipus legend but had no access to its psychodynamic interpretation.

From the start, Christoph's GAI sessions took shape in other than textbook fashion. Because of his anxieties, he continued occasionally to open his eyes, making sure that everything around him had remained unchanged. In so doing, he never lost the thread of what he was saying and subsequently picked up where he had left off. Christoph did not or could not follow direct urgings, for example, to choose the meadow as his topic. Other matters interrupted, which were connected with his immediate difficulties. As time went on, we were able to discuss current problems as well as his dreams and remembrances from childhood, which I subsequently had him partly reinstate in GAI. However, he was accustomed to spontaneously break this off after approximately 5–10 minutes.

Christoph gradually developed an amazing openmindedness for psychodynamic connections. They arose on the basis of free associations with what he had seen just previously. Thus, it became more and more possible to bring his anxieties under control and to put them into a context that he could understand. In some of his imaginings, which he sketched following the GAI sessions, one can still recognize the dynamics of what preceded both in the representation and in the sketchlines. The patient's fears were expressed in a picture with butcher's cleavers surrounding him (butchers were among his ancestors). Temporarily, during GAI, his hands and even his head were severed in this way. In a kind of birth fantasy, he appeared: "I made my way through a curved rubber pipe that was like a washboard on the inside." Po-

Special Features of GAI with Children and Adolescents 57

tency fears were documented in figures resembling spermatozoa, to which likewise dramatic and at the same time strangely magical things happened in GAI (Figure 5a/b).

Working through the compulsive fantasies and fears in GAI relieved Christoph to the extent that not only did he seem normal in his outward behavior to a great extent, but a few years after terminating therapy, he married a vivacious woman from a southern European country.

In Chapter 7, I will continue discussing the possibility of pushing through to infantile anxieties in GAI.

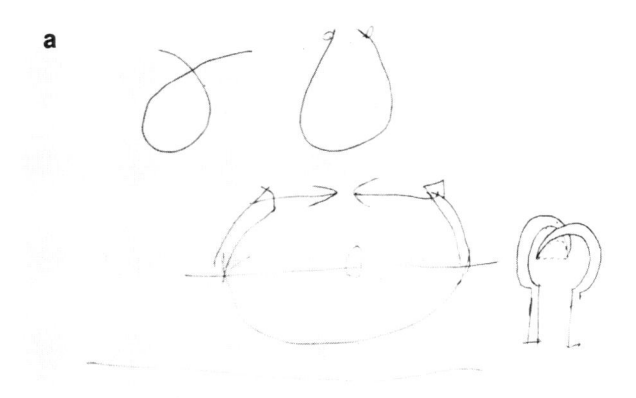

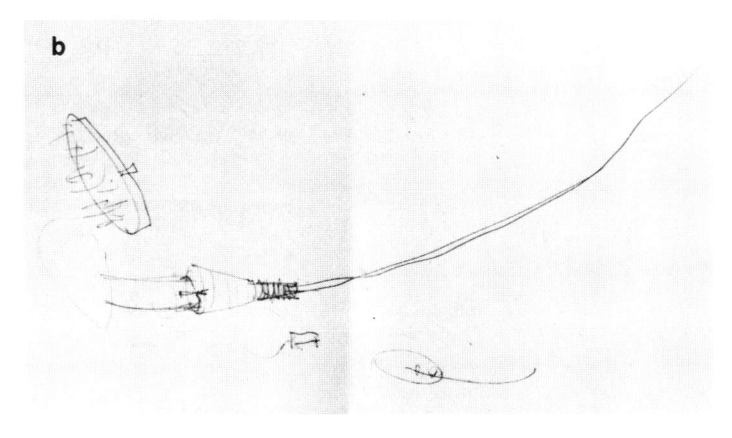

Figure 5a and 5b: Symbolic castration (butcher's cleavers) in an 18-year-old high-school student with "potency fears": (a) representations produced immediately after GAI; (b) dismemberment of a phallus-like figure.

This presentation concludes here. Its function was to supplement Leuner's basic statements by presenting detailed observations from treatments of children and adolescents with GAI and to draw certain conclusions. This presentation was consciously relatively unsystematic, to allow, as it were, the colorful variety offered to the GAI therapist by these age groups to shine through. Nevertheless, the choice of cases shows the diversity of the procedure and its applicability to children and adolescents. Furthermore, it seemed important to me to illustrate that the direction of therapy must proceed flexibly, often carried along by the therapist's intuitive and creative response to symbolically signaled conflicts. The therapy should be directed at the solutions, which the child frequently begins to hint at, and also toward well-considered, cautious offers for overcoming conflicts. This is done in the attempt to replace the presentation of contents of the "neurotic principle" with those of the advancing "therapeutic principle" (Leuner, 29). This task can, in the last analysis, be solved only by a specific therapeutic stance of the GAI practitioner, in the sense of an empathic and accommodating attitude that gives the patient full scope, on the one hand, and positive stimulation, adapted to the individual readiness of the child, on the other hand.

As is shown in a few of my cases, in which short-term crisis intervention was successful, GAI therapy not only is promising for this age group but also offers the therapist a high degree of satisfaction. In addition, there is the creative and aesthetic element, which appears suddenly and often naively in almost every therapy with these age groups and can frequently be incorporated into the psychodynamic concept in amazing fashion.

In the following chapters of this book, it will become clear that the GAI contents and the dynamic developments, as I have demonstrated them, are not specific to the therapist but occur repeatedly in similar fashion.

4

Guided Affective Imagery as Used in Diagnosis in Child Guidance

GÜNTHER HORN

Introduction

Within the framework of the Central Continuing Education Seminars in GAI in 1974, a number of impressive treatments using GAI with children 11 years old and older were presented. In contrast, the discussion of GAI with younger patients took up comparatively little space, although very good treatment success with younger children had been pointed out. At the same time, in a discussion involving some colleagues who had had previous training in other methods of child therapy, I was able to learn that they were fundamentally doubtful about the applicability of the daydreaming technique with children. They had even had difficulties in conducting a single session.

Within the scope of my position at the child guidance center of the city of Karlsruhe, I had at that time already had positive experiences in treating younger children with GAI. Nevertheless, I had the impression that I still did not have an adequate overview of the general applicability of GAI for the age group of 7- to 11-year-olds. Therefore, I decided to perform, in addition to our usual diagnostic procedures, a diagnostic GAI on all children in this age group who were referred to me for psychological testing. In two years, I have tested 89 children. Although the experiment is not yet completed, the results that are already becoming evident may be of general interest.

Survey

Let us begin with the most important finding: I was unable to perform a diagnostic GAI with only 3 of the 89 children tested. These children had all just turned 7, were severely disturbed in their behavior and could also be tested in other ways to only a limited degree. Diagnostic sessions could be undertaken with the remaining 86 children. The goal was to achieve a period of imagining, according to age, of up to 18 minutes. The following items were observed during the first and only test:

	Number of Children
Up to 4 minutes	2
5–7 minutes	9
8–12 minutes	59
13–18 minutes	16

The two children whose session lasted up to only four minutes were 7 and 8 years old and were very severely behaviorally disturbed. They had only limited intelligence.

I am aware of the fact that the real difficulties in the use of GAI usually emerge only within the scope of a more lengthy treatment. Nevertheless, the following outcome is surprising. In the group of 7- to 11-year-olds, it was possible to use GAI in the test phase in approximately 95% of the cases, that is, without special preparation and without the existence of a positive emotional relationship with the therapist, the basis required to support a therapy. To illustrate the contents of the GAI of these children, I would like to cite a number of brief segments from individual sessions in rapid succession. In so doing, I am consciously neglecting details and image symbols.

First, a brief technical comment about the performance of the tests is called for, since the percentage of children with whom GAI can be used successfully naturally depends essentially on the procedural technique.

What Were the Technical Conditions?

First of all, I would like to mention the therapist's personal contact with the child. In each case, an examination lasting approximately one hour preceded the GAI session. During this time, test procedures, such as Raven, the "family in animals," and scenotests were applied. The session in which the child discussed his worries and joys took on special status during this period. At the same time, we avoided distracting the child with the play materials that were often available in the examination room. The GAI session itself took place in another, somewhat quieter and slightly darkened room. Here it seemed to be very important that the curtains be already drawn when the child entered the room. In the child's presence, this act could perhaps arouse uneasiness. The curtains were never completely drawn, so that this special situation could still seem relatively normal to the young patients.

During the session, the children sat in an armchair with a high headrest, in which they could comfortably lean their heads back. From the few attempts that were carried out with the patients lying down, one can presume that it is generally better to suggest a sitting position. In addition, it is important for the *therapist* to sit *parallel to the child*, and not opposite him. The therapist's and the child's line of vision should face toward the dark side of the room rather than toward the window.

Another important condition seems to me to be preparing the child for GAI by providing him with reasons for it that he can understand. I usually ask whether he is already familiar with the "game of fantasy played with the eyes closed." Naturally, I always received the answer "no." By doing this, I have already aroused the child's curiosity. I told the somewhat older children that I wanted to do an experiment with them, in which they could imagine something in their fantasy. The following tape transcription reproduces a glimpse into the actual situation.

Therapist: ("I would like to run a test with you, in which you'll close your eyes for about ten minutes when I tell you to")—"Mhm."—("And I'll tell you what you should imagine and then you tell me what it looks like, OK?")—"Yes."—("Let's try it, OK? I'd like to record it on tape, so that we can listen to it again afterwards. Is that all right with you?")—"OK. Is the tape already running?"—("Yes, it's recording now. But I'll tell you exactly what to do.")—"Should I imagine I've got a hotdog?"—("Yes, you can if you want to. But first I'd like to tell you what you should imagine, and then you keep your eyes closed, until I tell you to open them again after about ten minutes.")—("So make yourself comfortable now, very comfortable. You can stretch out your legs in front of you and let your arms rest against the chair.

Close your eyes, you're very tired, and just enjoy sitting there. You forget everything around you and only hear what I tell you, and you're very relaxed and tired. You're breathing calmly now. Your arms and legs feel pleasantly heavy and warm. You're becoming calmer and calmer, and now you imagine you're in a meadow. And as soon as you can imagine that tell me what it looks like there.")—"Hm—it looks like—there's a nice little brook and people next to it! They're camping, we know them and they're cooking dinner and they invite us."

A further essential condition for GAI with children seems to be the pitch of the voice in which the therapist speaks. When I listened to the taped sessions again afterward, I found to my surprise that the pitch of the voice was very different in each case. With children who had a serious and reserved voice, I also expressed myself seriously and with reserve; with glad voices, mine was also glad. It seems to me that in this way, the child experiences a confirmation of his emotional state, at least to a greater extent than would be the case with the unchanging-neutral tone of voice of a diagnostician trying very hard to achieve objectivity. However, I tried not to let myself be influenced by two particular feelings of the children: by anxiety and by rage (or aggressive impulses in general). On the basis of the experiment with the diagnostic GAI, I came to the conclusion that with children, even more than with adults, it is necessary *to gear oneself to the patient in a very individual way*. I frequently observed, for example, that motifs that I introduced into the GAI session easily disturbed the flow of images and seemed rather to put the child under pressure to achieve. For that reason, I usually used only the motif of the edge of the woods in addition to the introductory motif of the meadow. In my opinion, the image of the meadow represents the greatest triggering stimulus for children of both sexes. The flower test, which is more meaningful for adults, seems to be less suitable for neurotically disturbed children. They seem to have greater difficulties than adults in observing and describing the details necessary for this test. Moreover, the image of a flower seems to be relatively uninteresting for boys in the latency phase. The equivalent for them would seem rather to be the motif of the tree. In fact, they frequently do picture trees in the meadow, which they then proceed to climb. The "edge of the woods" motif seems to be more suitable for the "first GAI" than the motifs "pursuit of the course of a brook" or "climbing a mountain." This motif allows the children to more easily develop a plastic image, and they are less tempted to evade this development by describing things they have actually done. Moreover, with children, the flow of images is often so autonomous and so spontaneous that new standard motifs sometimes do not produce any additional information at all. The associative procedure described by Leuner (35) for the intermediate level of GAI in the sense of associating in images takes precedence in GAI. The basic motif or the basic conflict then often appears in still other variations. An example is the case of an 8-year-old boy who first saw birds in the meadow. Then, birds came flying out of the woods. When he was then asked to imagine that he was lying, sleeping and dreaming in the meadow, he imagined birds again. However, it does not seem to be meaningful to totally dispense with presenting the familiar standard motifs, if need be, for the sole reason of being able to determine whether and to what extent the child is capable of picking up on image suggestions.

To illustrate the "diagnostic GAI," I would like to cite six short examples to show what can "emerge" from the imaginations of children. For the sake of comparison, I have chosen the motif of the edge of the woods which is often very productive.

Volker, 9 years, 6 months; Volker suffers from mild stuttering, a tendency to be constipated, and feelings of nausea, and has to masturbate compulsively in school.

Anamnesis: phimotic operation at 5 years of age. He slept in his parents' room until he was 6 years old. After suggesting the "edge of the woods" motif, he imagined, " . . . suddenly there is a rustling there in the midst of the bushes—and I—and in the grass—through the leaves. Suddenly I see—way in back—a wild boar."—("Mhm.")—"But—but—I hear a second noise, and there I see—suddenly—a snake wriggling there."—("Mhm.")

Beate, 9 years, 5 months; failure to achieve despite high intelligence, intense anxieties, would like to be a baby.

"A squirrel, a chicken, and then a fox—" ("And what do the animals look like?")—"The squirrel totally frightened—the chicken, it, it is upset, it's frightened, too, and and the, the fox, he's already licking his lips, because he would like to have the chicken."

Lutz, 7 years old; daily enuresis, excessive aggression, nailbiting. Severely disturbed relationship to his parents.

"A deer."—("A deer? What does it look like?")—"Whish!"—("What does it look like then?")—"Very dark."—("Can you describe it in more detail?")—"Very, very dark."

Andrea, 11 years, 7 months; intense school phobia despite good achievement and intelligence, headaches in her mother's absence, nailbiting.
"Well I see, that—a man appears, dressed all in black—and then he says stop, right where you are!"—("Hmh")—"I think, you can use the you-familiar form in speaking to me. And then he said, no dumb replies! Do you have any money with you?—I said, No! Crap! Am I in trouble again!"—("Hmh.")—"Then I'd at least like to have your pretty shoes for my little daughter! I said, Huh? I am keeping the shoes! Then I walked right away. But I was full of fear and kept turning around, the man just laughed loudly."

Ingrid, 7 years, 1 month; severe finger-sucking, nibbling between meals, defiant behavior. Striking in her anamnesis: several foster families and frequent illnesses.
"Here comes—a hare leaping."—("Toward you?")—"Yes."—("Hmh. Do you know what he wants?")—"He's hungry."—("Oh, would you like to give him something to eat?")—"Yes."—("Then, give him something.")—"A carrot."—("Hmh, does he like it?")—"Yes, I'll give him another one."—("Does he eat well?")—"Yes."

Peter, 9 years, 5 months; intense timidity, tendency to vomit, eating disturbances. Both parents, particularly the mother, are very depressive and overly anxious regarding the son's upbringing.
"A ghost that comes out at night, something like that."—("Look at it closely!")—"Now I see something, now a head is poking through, it's a skeleton in a dirndl dress."—("Hmh. Now tell me, what are you feeling while looking at it? How are you now?")—"Now—now I'm trembling, I'm afraid—it's flying toward me."—("What would you like to do?")—"Well, most of all I'd like to climb up a tree now, na na na."—("Look right at it! If you look right at it, nothing can happen to you!")—"I am already looking at it but when I do it stands still."—("Huh.")—"Then I think, I want to—the best thing to do is to disappear under the earth."—("Would you like to disappear?")—"Yes, if it would only go away! Yes, but then I run in back of it, but then when I'm standing in front of the tent, another ghost like it is standing in front of me."—("Hmh.")—"Then I want to go in back, and there's one standing in back. And then I'm—surrounded by ghosts like that."—("Whom would you like to have with you now?")—"With me? Strong people."—("For instance?")—"My daddy."—("Look around you! He's there!")—"Yes, mhm—he's—he's getting other people, they're coming with clubs, axes, and things like that and all of them are charging into the things—then."—("What does it look like now?")—"Something like a battle. I'm standing in the middle, and all around the outside they're beating each other up."

Use of GAI in Diagnosis

A statistical comparison concerning everything that came out of the woods was interesting to me. In 136 statements, there were named:

    20 × deer
    19 × squirrel
    18 × hare, bird, human form
     6 × dog, wolf
     5 × mouse
     4 × stag, fox
     3 × wild boar
     2 × weasel, snake, hedgehog, bear, horse, cat
     1 × lion, chicken, ladybug

However, to me, the question seems more important whether and to what extent difficulties arise during the performance of the diagnostic GAI with this age group. For this reason, the sessions were evaluated according to various criteria. The first criterion was the child's ability to keep his eyes closed. Within the framework of the Willinger seminars of the AGKB (German Association for GAI), the question repeatedly had been discussed whether and to what extent children should or can keep their eyes closed during the sessions. Seventy-eight children were observed (8 of the 86 children were not observed in this aspect, since this criterion was only incorporated into the program afterward). During the session, the eyes of the 78 were:

Closed from beginning to end and without tensing	32
Opened approximately one to three times, but then left closed	24
Closed with signs of tensing	6
Open part of the time	11
Always or predominantly open	5

The segment of the GAI of an 8-year-old girl demonstrates that imaginative productivity in GAI does not necessarily have to be adversely affected by the eyes being open. She had been referred to the center because of anxieties at night, intensely obstinate behavior, and slightly compulsive tendencies. The mother, who had been divorced, was studying pedagogy. She had little time for her daughter. The girl pictured, among other things:

" . . . Oh, yes, the conference is already starting. Now I really have to hurry."—("Mhm.")—"I'm still sitting on the grass and describing the meadow, so—first I have to get properly dressed."—("Hm. Well? What are you going to put on then?")—"I'm going to put on—a white—a, a white—I'll put on a gray tailcoat and—and a gray necktie and."—("Hm.")—"And I comb my hair and now I have to get into the car quickly. Oh no, I've still forgotten the leaves for the Stone Age—brr."—("You're in a big hurry?")—"Brrr."—("Where are you driving now?")—"Help, the tank is already empty. Now I have to get towed too."—("Ah, won't it go any further?")—"Hello! Can you tell me where the next gas station is?"

And in the discussion afterward: "Well, if I'm honest, I didn't have my eyes shut, honestly not."—("Were those things that you had already seen somewhere before?")—"Some of the things I'd seen before, but others I hadn't, but I'd already heard about them."—("And did somebody have to go to a conference before?")—"Yes, my mother."

One suspects that children who suffer from fear of the night or of darkness try, in general, to avoid closing their eyes. We did not run a control test with GAI in which all of the children had their eyes open.

Keeping their bodies calm and relaxed proved to be more problematic than the criterion of keeping their eyes closed, which approximately 75% of the young patients were able to do without great difficulty. The sessions were observed according to this criterion, too. Far more than half of the children were restless to extremely restless during the entire session.

The results, in detail, were as follows:

Behavior:

Extremely restless or tense	10
Predominantly restless or tense	41
Gradually calmer and relaxed	22
Quickly and deeply relaxed	13
	86

In considering these figures, one must not overlook the fact that all of these children, without exception, were referred because of pronounced behavioral disturbances. The young patient's ability to relax is not only a prerequisite for the general ability to perform GAI but also allows certain inferences about the child's general internal tension.

Thus, a clear correlation between the "motor unrest" that the parents reported in their children and the unrest that emerged during the GAI session could be established (see Table 1).

Only a small number of the children for whom complaints of the symptom "motor unrest" were reported attained a gradual or quick relaxation in the diagnostic GAI. Yet, each second child who had been presented without the symptom of motor unrest could relax well. Finally, the evaluation of criteria that directly influence the production of fantasies is of still greater interest than the criteria of closed eyes and body relaxation. Initiative is one of them. In almost all case discussions, the subject of the therapist's activity was brought up at some point. A control run by the therapist himself using comparative examination seems to be particularly important for this reason. The tape recordings of the cases I tested were subjected to critical inspection relative to statement intensity and the child's spontaneity in relation to the therapist's activity.

According to the respective overall impression, the initiative came:

	Number of cases
Exclusively from the therapist	9
Predominantly from the therapist	29
From the therapist as well as from the patient	24
Predominantly from the patient	24
	86

TABLE 1. DEGREE OF MOTOR UNREST

	BEHAVIOR DURING GAI			
	RESTLESS TO EXTREMELY RESTLESS AND TENSE	GRADUALLY CALM AND RELAXED	QUICKLY AND DEEPLY RELAXED	TOTAL
Number of children with the symptom "motor unrest"	19	2	1	22
Number of children without the symptom "motor unrest"	32	20	12	64
	51	22	13	86

Insufficient initiative on the part of the patient in GAI seems to represent a contraindication for using this procedure only when it occurs in extreme form.

Here is an example of positive therapeutic results despite modest progress on the image level. I have an 11-year-old boy in treatment whose initiative in the affective imagery is considerably hampered. Nevertheless, his behavior has changed to a great extent since the start of treatment (35 sessions). His unusually severe difficulties in adjustment in school and in the family surroundings are now scarcely evident. His extreme aggressions have disappeared; stomach trouble and thumbsucking no longer are apparent. What had happened during these 35 sessions? First of all, the boy ended each GAI session by lying down to sleep in the meadow. In later sessions, he regularly returned home to his mother at the end. This is a sign that strong regressive tendencies were appearing in the imagery, which becomes comprehensible when one knows that he suffered greatly from the actual separation of his parents. Typically, these tendencies had already been often observed in his play therapy sessions. In the meantime, this tendency has subsided considerably. However, the house is, now as ever, often the focus of his fantasies. Once he painted it: gray, empty, and without animation. In the thirty-first session, I tried more intensely than otherwise to make him conscious of his attitude of expectation. Unusually long pauses arose:

" . . . One house has a picture painted on the side of it—a meadow with a brook and a few flowers. There is a small castle in between."—("Mhm.")—(40-second pause)—"The gray house has no bars in front of the windows. It has a projecting roof and is about three stories high—"—("Hm. Is there a picture on it, too?")—"No, it's gray all the way around, the windows are white—the frame is totally white, on the roof there's a German flag."—("Hm.")—(55-second pause)—"Now I'm going back to the meadow again, go to the far end again."—(18 seconds)—("Wolfram, is it more comfortable for you, if I ask you questions more often? Or would you like . . . ")—"Yes."—(" . . . to tell it all on your own?—How does it seem more comfortable to you?")—"With questions."—("Did you like them?")—"Yes."—("What was nice about it?")—"The pictures."—("Did you have any idea if someone lived there?")—"Yes, of course."—("Did you see anyone?")—"Yes, in the kitchen, a woman was working."

Wolfram never had a father who took him by the hand and showed him the world. And his mother, employed and overburdened with the care of five children, had never been able to give him much attention either. Consequently, one can understand the boy's wanting to be "taken by the hand" more than other children in the internal imaginal world do. In the last GAI session he moved into his own unoccupied gardenhouse and furnished it domestically.

We would like to assume that deficient feelings of security and an absence of attention on the part of the environment appear not only

USE OF GAI IN DIAGNOSIS

in the contents (for example, in the symbol of gray houses,) but also in the scantiness of utterances in GAI. Presumably, one can come to a further conclusion: even when children's imaginal contents appear depressing and threatening, merely the expressive capacity shown by them can be an important indication of a good prognosis for treatment by GAI.

The sessions were additionally evaluated according to the criterion of aggression. In the GAI contents of adults and adolescents, the menace that a patient experiences during his fantasies is often regarded as danger, extreme threat, or, as the case may be, even as a contraindication for treatment using GAI. For this reason, I was especially interested in whether and to what extent children also feel threatened in the GAI session and whether they spontaneously imagine aggressive wishes.

In the first GAI, the immediate threat recedes for the child:

Felt extremely threatened	0
Felt more-or-less threatened	27
Felt neither threatened nor induced to own aggressive wishes	50
Some aggressive wishes imagined	9

None of the children suffered an "imaginal death." However, through two examples I would like to show how aggressively or threateningly experienced fantasies can appear; one with externally directed aggressions, another with autoaggressions:

Rainer, 9 years, 7 months; lying, stealing money from home (buys capguns with it), truancy. He imagined among other things: " . . . I have a really large plane—and I fly away with it—to—an unknown place—and there I am transformed into—some kind of—an animal—a tiger, for example, and then go hunting . . . "—("What's it like to be a tiger?")—"Good, then one has a good feeling—one can jump a long ways—and then I chase after—a kangaroo—but I don't catch it, it's too fast, but then I get another catch, a bird . . . and kill a bull, a kind of buffalo—and then I eat it . . . I change back into a person, and that's enough till the evening."—("Hmh.")—"But—in the morning, at noon, and in the evening I change myself into a cheetah, and afterward, when I'm hungry, I eat what I've caught."

Clemens, 9 years, 3 months: enuresis and encopresis during the day and at night, failure to achieve in elementary school despite above-average intelligence, rocking his body back and forth at night from early childhood on. The environmental situation has been burdened for years, particularly by the severe marital problems of the parents. The boy imagined, among other things: "Dream of—a crocodile, that enters—it wants to eat me up—"—("Tell me what the crocodile looks like?")—"It's long and big—sharp teeth—it's green."—("And where is it, in what area now?")—"In the children's room."—("And where does it enter?")—"From the door."—("Is the door open?")—"Yes."—("And what are you feeling when it comes in?")—"I was afraid and—quick under the pillow—blankets—got out—and got out my—gun belt, that I had under the pillow—and threw at him—"—("And what's it doing now?")—"It went out again."

In general, I found that one can more easily intervene in the threatening experiences in GAI of children up to 11 years of age than in those of adolescents and adults.

Considering the sessions according to the criterion of reconciliation with hostilely experienced objects was likewise interesting and yielded the following results:

	Number
No reduction in distance	62
Partial or total reconciliation	24

With that, we have reached the point of considering the therapeutic effect of the GAI sessions. The assessment of the sessions according to the aspect of the developmental tendency gives clearer information about this. A change for the better in the imagined scene took place within the diagnostic GAI of 38 of the total of 86 children. According to Leuner's (40) observations, such changes in GAI signal therapeutic progress even when they are not yet apparent in the outward behavior of the patient.

Michael, 7 years, 8 months; suffered from intense anxieties and was severely inhibited aggressively. His capability to play was considerably impaired. The contents of GAI were very scanty and led one to conclude that he had little initiative. It rained in the meadow (sign of a depressive mood). Eventually, Michael was threatened by a wolf that wanted to eat him up. It was already licking his feet. Michael fled to a nearby apple tree. Following my advice, he now fed the animal according to the management model of feeding and reconciliation (Leuner, 31). Afterward, the wolf left satisfied. When Michael returned to the starting point in the meadow after a while, the sun was shining. "Now the grass is all different, green." (Positive transformation phenomenon, see above.)

Christian, who was 8 years old, behaved similarly. He also suffered from intense anxieties. His relationship to his mother was severely disturbed. At first, it was raining in his meadow, too. Cars drove through puddles and splashed him. He did not have an umbrella along. Later, however, he picked flowers and put them on a table at home. When he imagined the meadow again, it was animated. Now, the sun was shining and the garden was blossoming.

From these cases it becomes clear that even in the short first session of GAI, such positively accented changes in the imaginally experienced emotional situation of the children can take place. They can be verbalized as "hope, cheerfulness, enrichment, expansion of personal possibilities," and so forth. On questioning the children afterward, they responded almost without exception that the fantasy game was "not difficult."

However, it was impressive that almost all of the young patients left the impression of being decidedly relaxed and relieved after this one session and that this observation was also later confirmed by many parents.

The tests described here using the diagnostic GAI with 89 children ranging from 7 to 11 years of age were absolutely positive in 95% of the cases. Even though this result does not prove that a systematic treatment of GAI with the same number of subjects would take a positive course, it does become recognizable that introducing the GAI session is not a problem with this age group, if the technical requirements described are met. Not infrequently, we begin to see an obvious therapeutic effect even in the first GAI session.

Our experiments are not yet concluded. They are to be continued until finer age-specific differentiations can be made. Our findings invalidate, in any case, the skepticism mentioned at the beginning, that GAI can not be used at all or only to a very limited degree with children in the age range of 7–11 years. It has been generally found that applying GAI promises a considerable reduction in time for a conflict-centered (psychodynamically oriented) psychotherapy. This finding, which we have taken from the first GAI as presented here, is to be further exemplified by the course and results of therapies.

II

THERAPEUTIC RESULTS

5

Outpatient Psychotherapy of Anorexia Nervosa Using Guided Affective Imagery*

EDDA KLESSMANN AND HORST-ALFRED KLESSMANN

*Revised edition of the original publication as it appeared in *Zeitschrift für psychosomatische Medizin*, 21 (1975) 53 and in *Handbuch der Kinderpsychotherapie*, Ergänzungsband, p. 296 (1976).

Within the framework of this chapter we do not intend to make a detailed analysis of the symptoms and problems of treating anorexia nervosa (AN). Our concern is, rather, to point out the possibilities for outpatient therapy and to allow the psychodynamic background of what is occurring to become transparent, as far as this background is represented in the imaginations of female patients during guided affective imagery (GAI).

First we will briefly sketch our outpatient procedure. As husband and wife (internist and pediatrician, both psychotherapists), we have developed since 1969 a carefully constructed program. In this program, both the somatic and the psychodynamic backgrounds of the illness are taken into consideration. Here, we assess 20 of our own cases (some of whom were previously in hospital care without success) who have succeeded in overcoming the "self-reinforcing circle" (56) of anorexia nervosa. Certain behavioral features in the area of eating and disturbances in the perception of self-esteem were partially retained after outward normalization. Since our procedure is a short-term therapy, an outpatient aftertreatment (for instance, in an analytic group) was recommended for some patients for further treatment of their, in our opinion, severe, fundamental narcissistic disorder.

Fleck (12), among others, has pointed out that "interpersonal relationships remained persistently disturbed with few exceptions, if a longer-term treatment was not possible after the disappearance of the threatening symptoms."

The Technique of the Two-Track Outpatient Treatment

Internist's Procedure (Horst-Alfred Klessmann)

Besides examination and supervision and the treatment of intercurrent somatic illnesses (which, by the way, rarely occurred), the internist's most urgent task proved to be intervening in the problem of and recouping the severe weight loss of the patients (around 30% of the mostly normal original weight).

A stomach tube (13) inserted for several weeks' duration could, of course, not be used on an outpatient basis. For that reason, intravenous infusions were administered in the office. In general, 1 or 2 infusions per week and, altogether, no more than 10 infusions per patient proved to be sufficient. We preferred an infusion solution of the plasma expander type in order to avoid additional electrolyte displacements. In addition, Biosorbin ®, a fully balanced nutritional preparation administered orally, was prescribed over a longer period of time. Surprisingly, we were successful in getting the patients to take the medication fairly regularly with the argument that this calorie supply would be "burned up" primarily for energetic purposes.

Although we dispensed with weighing our first patient regularly (we were afraid she would discontinue treatment) weight controls were subsequently built into the therapy program from the start, and the patients were left no other choice. To prevent deceptions, the patients were weighed undressed (nevertheless, they tried to "cheat" by drinking plenty of water beforehand). The appointment for the next infusion or further medication was then set according to the weight level.

The infusions by no means proved to be "just" a somatic therapy. They had a distinct psychotherapeutic side effect, which we consciously incorporated into the overall strategy. The patients experienced the infusions very ambivalently, in part as a reward and/or a masochistic criminal proceeding. During the infusions, their unconscious fantasies of rape were very clearly documented in the patients' statements and were dealt with in the ensuing psychotherapy sessions. (In Chapter 7, we have shown that for adolescent drug users, the intravenous injection, the first time they "shoot up," can have the significance of an initiation rite!) The illusion of (seeming) independence, so important for anorectic patients (fantasies of omnipotence to defend against dependency needs) was taken away during the infusions, since the patients could not act independently at those times but had to let something unavoidable occur (which could not be changed again afterward either). Because of poor circulation, and in part also because of a certain restlessness during the approximately two-hour-long infusion, there were occasional paravenous infiltrations. Such swellings on the inner side of the elbow, usually negligible, were commented on with considerable alarm: "I have the feeling something foreign, strange has intruded there. I would have liked to scratch it out right away"; "I feel I've been punched full of holes." The patients' unreal relationship to their own bodies (8, 17), distorted by unconscious fantasies, became transparent here. However, those patients who subsequently recognized that these anxieties were unjustified seemed clearly to us to have more "ego strength." Battegay (3) correctly says of patients experiencing anxiety, "Only when they can also learn to endure anxiety and to conquer it will they gain in strength."

The amenorrhea, which initially persisted even after the normal weight level had been reached, was treated with a hormone-free phytotherapeutic drug, and if necessary, the patient was sent to a gynecologist for subsequent hormone therapy.

Interestingly, the patients almost never asked to have laxatives prescribed. Some of the patients obtained them without having a prescription, or they were able to induce a bowel movement in a way that seemed satisfactory to them by eating large amounts of raw fruits and vegetables (at times, they ate nothing else).

Generally, the internist assumed a matter-of-fact, decisive, firm, and single-minded attitude toward the patients. He made it clear that if they did not keep to the suggested course, he would consider assignment to a hospital unavoidable. As outpatients, however, the patients would have the distinctive opportunity of contributing to the treatment.

Psychotherapeutic Procedure and Particular Experiences Gained with Guided Affective Imagery (Edda Klessmann)

The psychotherapeutic treatment proved to be particularly productive, when GAI managed to open "the gates to imaginal consciousness" (40). The sometimes amazingly rapid unburdening, the easy maneuverability, and the direct vividness of the inductively produced daydreaming technique developed for the anorectic patients with their mostly lively imaginations, (to some extent,) into the instrument through which the psychodynamic dialogue not only started quickly but also could be continued even through phases of resistance. Considering the poor physical condition of our patients, the rapid progress of therapy was important; we were dependent on being able to gain and deal with "material" as quickly as possible. On the other hand, we were aware of the danger of too suddenly breaking through the anxiety-laden defensive front of the radically changing pubertal phase (14). When one is well acquainted with the GAI technique, one generally senses whether it is still too soon to release defenses. The patients display accompanying psychosomatic reactions when caution is called for. As a rule, however, the patients do not produce more imaginative material than can be dealt with.

With our patients the psychotherapeutic relationship proved to be a complicated balancing act. The first step was to produce a positive transference, capable of enduring stress, in order to overcome negativistic defenses and mistrust. Kohut (24) commented that "the question of the therapist's activity [is] of great importance in the treatment of certain specific types of narcissistic personalities" and called to mind Aichhorn's (1) active technique with neglected adolescents. A similar "start" also proved successful with our patients. Later on, there were occasional phases of "identity confusion" during which the therapist simply had to "maintain understanding and affection for the patients without either devouring them or offering himself as a totem-meal" (11).* The "twin, mirror, and idealizing transferences" that Kohut (24) described also played a part during extended stretches of therapy. In the GAI sessions the different transference relationships were expressed, in part, in symbolic comparisons ("evil witch," "friendly female farmer distributing apples," "fantastic horsewoman," etc.).

In addition to the regressive transference mechanisms, prospective possibilities for identification with the therapist began to func-

---

*Compare, in this connection, the descriptions of Patient A at the end of the first GAI quote.

tion. In the past, the patients had not been able to satisfactorily complete the "process of defining limits and of reconciliation" with their mothers. In the psychotherapeutic relationship it became easier for them to find the latitude necessary for their own identity.

The number of therapy hours ordinarily totaled between 30 and 60 (including, for instance, a possible relapse) so that one can speak of ours as a short-term therapy. Three of our patients moved away because of courses of study and thus left therapy at a time that had been determined at the start. After good immediate results we can not yet assess the further development, but we can already say that for the difficult and lengthy treatment of anorexia nervosa GAI seems to offer optimal therapeutic possibilities.

Therapeutic Procedure with the Mothers of Anorectic Patients (Edda Klessmann)

An important additional factor in therapy was the work with the patients' mothers, whom we want to mention briefly. We found the often-quoted statement (51) confirmed that the grandmothers (or substitute grandmothers) strongly dominate in anorectic families. Thus, our patients' mothers seemed neither to have determined their own identity nor to have affirmed the female role. Often they believed that they had to justify themselves in an overcompensatory fashion (guilt feelings) and to "sacrifice" themselves. Apparently this role conflict was sensed by the really empathic daughters and led, during their puberty, to their own identity crises. Richter (44) wrote of the close "dialogical clinging" of the anorectic patients to their mothers. Therefore, it was necessary to help the mothers gain greater autonomy and to give them insight into their share in this pathological role conflict in order to facilitate the reciprocal process of detachment from the mother–child symbiosis.

CASE STUDIES

Case A

At the time of referral, A was 18 years old and had suffered from AN for one and-one-half years. At first, it had started—as usual—"harmlessly" and had developed progressively following an extended stay abroad. Her first physical contact with a boyfriend had been the triggering situation leading to intense sexual anxieties and fantasies of oral conception. Washing and compulsive rituals followed. In spite of factually correct explanations, she had the idea that "children and feces come out from behind." She perceived menstrual blood, discharge, and feces as "disgusting": "Somewhere in the belly there must be a deeper orifice that separates these from the rest of the body." During a gynecological examination she was embarrassed by the idea that the physician must have got excrement on himself.

At this point we would like to indicate, in particular, the early narcissistic disturbances of the "body self" (24). Kohut described "the lability of the self-image of adolescents . . . in whom the firmness and elasticity of the core of the self are strained too far" and referred to a behavior that we never found missing in our patients: intensive self-observation in mirrors.

In this connection, A reported, "I often check myself in front of the mirror. If I only see my belly, it's not so bad. But my whole figure? Then, I always have to slap and stroke my belly." Kohut's statement that the narcissistically disturbed person "tries to make up for the lack of narcissistic cathexis" by 'stimulation' of the entire body self (by forced bodily activity) was entirely confirmed by A, as she spoke, for example, of the "comforting feeling" after intensive bodily manipulation. Moreover, she stated, "When I eat something secretly in the evening (she had developed a ritual of eating raw foods at her bed), it annoys me incredibly if someone comes along. I seem to be totally at peace with myself and have the feeling that the others notice this. I feel so incredibly at the mercy of others, if, for instance, they laugh at me." (Here, Kohut's reference to the "intense feelings of shame" as an indication of a deeper disturbance in the area of the libidinous cathexis of the self.)

A had developed a strong ambivalence toward doors: "Sometimes I have the need to keep everything open, for example, to have a curtain instead of my door. Then I would not have the 'door feeling' any longer that someone could enter at any time and a change would occur again and again. On the other hand, I also need the door closed when I have to exist completely for myself."

Regressive symptoms of enuresis and encopresis appeared temporarily. In this phase, when the family was at the table, A had to leave the bathroom door open and, while defecating, had to communicate through the open door with the family, while they ate. She remembered that she had felt terribly lost as a small child when she had left something in the potty and, with her pants down, had gone looking for her mother to wipe her. According to her mother's description, A had not developed a phase of ego assertion.

Unlike her brothers and sisters, she had, according to her mother, "very precisely" but apparently ambivalently sensed at that time, that her mother felt overstrained in an extremely difficult situation. During that period, A had also stated that she would rather become a father; then one could lie on the couch and read books and would not need to work as much as the mother. She apparently rejected the actual maternal image, which did not seem to hold much attraction for her, and unconsciously longed to be back in the wishful world of symbiotic unity with her mother, as we learned from subsequent statements.

The striving for compensatory achievement and cleanliness, which the grandmother had taken over (pathologically severe superego mechanisms, 54, as well as "fusion tendencies," 24, with idealized objects), characterized our patient's frustrated attempts to attain an experience of self-esteem.

In GAI, A's disorder was impressively demonstrated to us.

In her imaginal pictures she first experienced a superficial world of "pretty appearances." Delicate grasses swayed in a pretty but sterile landscape. However, A soon fell into a deep regression and at times spoke in a scarcely audible voice. In the sixth GAI session she imagined a house. She did not know whether it was supposed to be her own or that of her parents. The house was situated high above the city, a "fantastic location." However, she could not go through the entrance. When she had finally gone in through the terrace, the emptiness of the house struck her. It became more cozy only after, in her imagination, she had managed to start a fire in the fireplace. Now, she said literally, "Suddenly, there's a soft carpet there, too. Now I'm reading a book. Colorful plates, apples and nuts are lying there. All of a sudden, my brothers and sisters and friends are there. It's like during Advent. My mother is baking cookies. It's like it used to be but too childish. Now I'm trying to move close to the others. My mother is coming out of the kitchen; it's really warm there, it smells good. For the first time I'm eating with the others completely uncompulsively and feel good doing it. But suddenly all of the others go away. All at once I'm alone again, so terribly alone. I don't know at all, anymore, where and how I am. Now, I go to my mother in the kitchen. Here I'm not alone any longer; instead, I'm her opposite. Earlier I became totally merged with the others, as if engulfed by them. Then, I was spit out again. That was as if we had mutual thoughts. When they are gone, I no longer have any confirmation. It's always like that. The others can simply say 'no.' I can't do that. I identify completely with the others. When they go away, it's as if they are cut away from me. When I meet them again, at first they're uncannily strange to me. Then, there's another similar identification, that is also as if I'm being engulfed, so that they aren't strange any longer. But when they go again, I don't feel well. Then I would like to give up everything again, spit them out" (problem of defining self-limits, of working through object loss).

In a later GAI session, the patient was encouraged to eat a few chocolates that she had imagined beforehand and offered to a young couple. In her imagination, she ate three chocolates with much inner aversion. Immediately afterward, she felt that the atmosphere had become "warmer" and she had come somewhat closer to the young couple. However, she now increasingly complained of an "uncanny inner feeling of disgust" and made movements as if she had to throw up the imagined chocolates again. At this point, she looked pale and strained (guilt-cathected oral engulfment; see also subsequent GAI experience with the maternal breast).

In the next GAI session, after having picked apples "in masses" and being urged to try just one, she chose "a rotten one with a worm in it." She was able to bite into this one because it was "worthless anyway"; she was able to throw it away then, too. Otherwise, she found it immensely difficult not to finish eating something once having started to eat it. "I cannot throw it away; at most, give it away. Then I have to engulf all of it into myself. If one portion remains outside, it is like a threat. I perceive it almost as destructive" (Selvinis's "bad object," 48).

We managed to work through the eating problems in the course of one winter (one psychotherapeutic double session per week, on the average) to the extent that the patient not only ate approximately normal amounts but could also partake of foods that had up to then been strongly taboo (bread). In the course of the summer, she had attained her normal weight and was menstruating again, though still complaining about "annoying residues," by which she particularly meant certain compulsive rituals during meals.

After a crisis in her relationship to her boyfriend, she had a relapse in the course of the next winter.

This time in GAI, she no longer returned to the described archaic ideas but to very early, real remembrances, which went back into the oral phase. She said, "I see myself as an eighteen-year-old with a bicycle, suddenly, however, to myself I seem like twelve. I pass by Baker D; there is an awfully good smell. I breathe in deeply, am very tense. Am very hungry for cherry cake, as if warmth or softness exuded from it. As if one—when one holds one's breath—sucks in—is at mother's breast. Now I have such a strange fear, as if someone is biting off the breast. I see the breast distinctly before me. I am afraid that it isn't there anymore, that someone is hurting mother. Now I see the piece of meat just like the earlier image with meat. It is terrible."

On the suggestion to suck only delicately, A replied, "That won't work, it's as if my neck is rigid. I always see the breast, and next to it the cake. I would like to bite into it." The patient is urged to try to make eye contact with the mother: "Yes, now it's getting better, now I can drink too. It's getting warm, close and soft."

Later, she stated the following, "I see myself crouching in front of my cupboard in the kitchen, in a position like an embryo. Only when I stand up, do I notice how long I've been sitting, because my legs hurt. I am undecided about eating, as if threads were holding me fast. Now I'm falling; it is as if cushions were there; I have no sense of time at all. I nibble first on cottage cheese, then cookies, and bite right into them, too, as if the mouth were biting along by itself, like an animal. Finally, I want to eat bread. Then, I don't feel so empty anymore."

At a later date, we had her imagine the mother's breast in GAI again. This time, she saw her younger brother at the breast and stated, "I'm afraid he could bite something off. Now I couldn't drink at the breast; I seem strange to myself. I'm sitting with my mother's arm around me and drinking out of a cup. I'm trying to get further onto my mother's lap; my sister pushes me away." When urged again to take her mother's breast, she said, "I'm a little bit afraid, as if I would bite into it. It won't work." The patient is encouraged to imagine taking the breast and giving it up again. She becomes noticeably restless and breathes heavily. Only after being encouraged to look for eye contact with the mother does she become calmer. When her mother buttons up her blouse, she has the feeling that the breast is gone, that she can never get it again. She carries out the request to unbutton and button her mother's blouse herself. This action relieves the patient visibly. She no longer feels "so helplessly at the mercy" of the situation. In her imagination she subsequently has her mother bathe her. She finds this "fantastically good." After emerging from GAI, she seems clearly relieved. Within the two months following this GAI session, a very strong improvement resulted.

In the third winter, another crisis in her relationship to her boyfriend led to a mobilization of the old anxieties and defense mechanisms. This time, the patient could cope without therapeutic help. Interestingly, she also lost "the annoying residues" when she felt she had been roughly confronted with her abnormal behavior in a politically active group, feeling at the same time supported by this collective. She "handled" a group confrontation and now was even ready to continue with aftertreatment in the form of group psychotherapy at her place of study. We regard these signs as a clear increase in ego strength.

Case B

Whereas with Patient A (at the beginning of treatment) the neurotic compulsive-depressive part of her personality became predominantly evident, Patient B tended to process experience in a predominantly hysterical fashion. With her, transitory episodes of bulimia (excessive intake of food) appeared more often, followed by vomiting. Leuner (31) has partially related aspects of GAI to behavior therapy, in particular as a technique of modifying behavior through confrontation. GAI could be performed in this way particularly well with Patient B.

The 17-year-old B—with a maximal weight loss of 40%—soon regressed deeply in GAI, too. The imaginal world she described had an "archetypal" symbolic quality.

In the third GAI session, she experienced herself in a role similar to that of the king's daughter in the fairy tale "The Frog King." In the sixth GAI session, she was confronted with the theme "moor." Spontaneously, she imagined herself in a pool, which was "terribly dirty and muddy." Nodding her head, she saw her reflection and commented, "My hair hangs down dirty. My face is smeared, too. I look like a little witch. The filth disgusts me. Most of all, I'd like to run away." However, the therapist encouraged her to hold onto her reflection and to accept it in this unidealized way, which she was able to do after exerting some effort; "Now it's getting better, I'm getting used to myself." For the first time, she clearly managed to master the anxiety of her eating problems. Only now did she tell about looking at herself in the mirror daily. She stated the following: "When I look at my legs from above, I think, what fat calves I have. But in the mirror I see that they are only sticks after all. It's just the same with my stomach, too. But with my face, it's just the opposite. In the mirror, it seems fat to me; when I touch it, it seems thin."

Until she was 12 years old, she had slept in the same room with her grandmother and always admired her as a "strong personality": "She only had to look at one and one was humiliated. When she died, it was as if I were empty. If I ever identified with any woman, then it was with her. Everybody bowed before her."

In the eleventh GAI session, spontaneous "flights of fantasy" appeared for the first time, symbolizing B's excessive level of aspiration. "The theme of grandiosity in the fantasy of flying" (24) enraptured her and made her anxious at the same time. Temporarily, she had the idea that she could not return to earth anymore. B, who also perceived this state to be like an ethereal swimming, stated anxiously, "Suddenly I have the feeling I can't go ahead or back either. Time is standing still. Everything must stay as it is." After a long pause of clear perplexity and uneasiness, she reported, "Now I'm looking at the clock. Suddenly I'm looking forward to having a task. I come back to earth, and all at once, I know exactly where I have to go." Then followed a vision of the future, concerning what she imagined her profession would be.

After the disillusioning of the "grandiose self" (24) and an acceptance of the reality principle was possible in the catathymic imagination, the patient became substantially more well balanced. She no longer had any eating problems, her weight was normal, her digestion was good, and her menstruation had returned. In the following winter, she experienced renewed identity problems, when she had difficulties being accepted into her program of studies. During this period she had pronounced "streaks of gorging herself" at home, which, however, did not lead to overweight or to renewed gauntness because of subsequent vomiting. In her diary she now revealed intense anxieties concerning loss of control. She described these as follows: "I simply cannot cope with my body, and therefore, I cannot tolerate caresses either, because I'm afraid that my own feelings and drives could overwhelm me."

After eight psychotherapy sessions (this time without an accompanying treatment by the internist and exclusively with GAI), the patient was again in a stabilized condition. After that, she got through her first semester of studies and was able to manage well in an intimate relationship with her partner. However, she still had the need to "gorge" herself when she felt internally tense. She considered a group aftertreatment where she was studying.

Case C

The 14-year-old female high-school student C was the only one of our patients who had lost her father at an early age. The mother spoke of a purely "female family" and said, "We three girls sleep in one room; grandma sleeps in the next room." The "three girls" were the mother, the patient, and her younger sister. This mother had not overcome the position of dependency on her own mother either.

In GAI, C often experienced the obstructive motif of a board, which came down in front of her face each time she wanted to pass through a door (Figures 1 and 2). For that reason, she could go through a door only backward or crawl through the doorway. In our opinion, she was thus symbolically expressing the not-yet-performable step over the threshold as an expression of the fear of becoming independent (see also the "door problem" of Patient A).

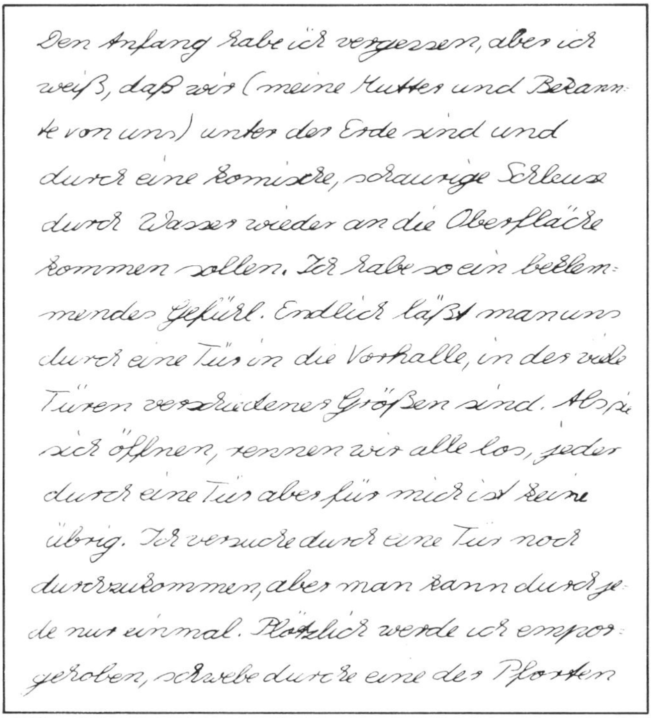
Den Anfang habe ich vergessen, aber ich
weiß, daß wir (meine Mutter und Bekann-
te von uns) unter der Erde sind und
durch eine komische, schaurige Schleuse
durch Wasser wieder an die Oberfläche
kommen sollen. Ich habe so ein beklem-
mendes Gefühl. Endlich läßt man uns
durch eine Tür in die Vorhalle, in der viele
Türen verschiedener Größen sind. Als sie
sich öffnen, rennen wir alle los, jeder
durch eine Tür aber für mich ist keine
übrig. Ich versuche durch eine Tür noch
durchzukommen, aber man kann durch je-
de nur einmal. Plötzlich werde ich empor-
gehoben, schwebe durch eine der Pforten

FIGURE 1: Patient C's initial dream, in which, after a symbolic act of birth, the problem of passage to puberty remains unmastered (in the dream she cannot get through the door). Instead, she had the following "narcissistic flight fantasy": "I have forgotten the beginning, but I know that we (my mother and acquaintances of ours) are underground and are supposed to come to the surface again by way of a strange, horrible drainpipe through water. I have such a suffocating feeling. Finally, we are allowed to pass through a door into the hall, in which there are many doors of varying sizes. When they open, we all start running, each one through a door, but there is none left for me. I try to get through a door, but one can go through each door only once. Suddenly I am lifted up, float through one of the gates."

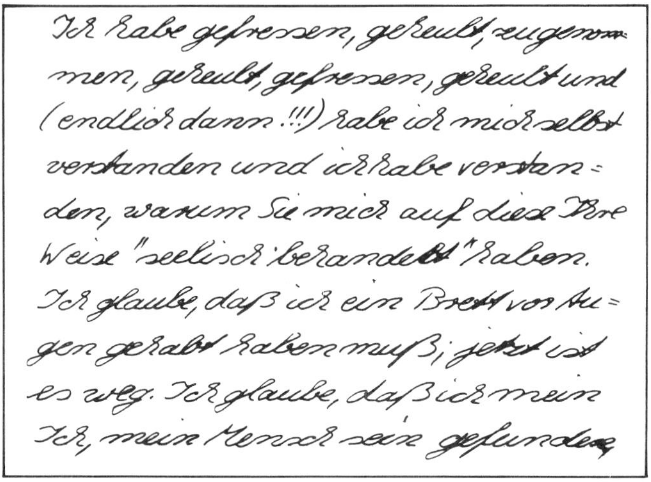
Ich habe gefressen, geheult, zugenom-
men, geheult, gefressen, geheult und
(endlich dann !!!) habe ich mich selbst
verstanden und ich habe verstanden, warum Sie mich auf diese Ihre
Weise "seelisch" behandelt haben.
Ich glaube, daß ich ein Brett vor Augen gehabt haben muß; jetzt ist
es weg. Ich glaube, daß ich mein
Ich, mein Mensch sein gefunden

FIGURE 2: Excerpt from a letter by Patient C after she had managed the cited problem of passage for the first time in GAI (freely passing through a door as an expression of her independent step to maturity): "I gorged myself, cried, gained weight, cried, gorged myself, cried and (then finally !!!) I understood myself and I understood why you have given me this, your 'psychological treatment.' I think that I must have had blinders on; now they're gone. I think I've found my ego, my human-ness."

In the practicing-confronting style of GAI already mentioned, C was encouraged to confront the door's obstacles. To her own surprise, she succeeded in doing this for the first time in the eleventh session.

She documented this breakthrough with an exclamation, both startled and delighted, "Huh, I've done it!" It was interesting that she was able to do this only after she had spontaneously imagined a second "ego," a girl who displayed the role of her own helplessness: "She looks so sad. She is lying there as if she were dead, a little bit twisted. I don't know, perhaps she ran so fast, ran away to somewhere."

Thus, in the alter-ego fantasy (24), C first assumed the role of the strong pioneer, then subsequently identified in an interplay alternately with the weak and then with the strong "twin." In the breakthrough described above, not only had she symbolically tested a new step in GAI, but she also gained weight after that session.

A subsequent dream, which took place after a discussion with girlfriends about "having children," showed us, nevertheless, how vulnerable her relationship to her "body-ego" remained for the time being. The patient stated, "I bled, under a blanket. I believe somewhere from the belly, the navel? Someone peeked through the door, laughed maliciously. I was terribly ashamed, I thought to myself, that could happen to that person, too!" It proved useful subsequently to focus again on such anxiety dreams in GAI and to work through them. We were, of course, cautious in giving interpreta-

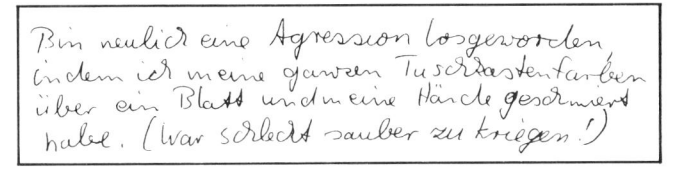

> Bin neulich eine Agression losgeworden,
> indem ich meine ganzen Tuschkastenfarben
> über ein Blatt und meine Hände geschmiert
> habe. (War schlecht sauber zu kriegen!)

FIGURE 3: Section from a letter during a vacation by the same patient (three months later). The relaxed handwriting and the natural style document the internal change: "Recently got rid of an aggression by smearing all my watercolor paints all over a piece of paper and my hands. (Was hard to get them clean!)"

tions to the 14-year-old patient, as we sensed that to do so would have unrolled the defensive front too early. Particularly here, GAI proved itself in its interpretation-free, "matter-of-course," and deeply effective possibilities of conflict mastery and personality development (Figure 3).

DISCUSSION

With the two-track psychosomatic therapy of anorexia nervosa, we believe we have found new possibilities for an outpatient treatment.

The motivation, particularly necessary for voluntary, outpatient therapy, which—as is well known—almost never exists in these patients (or is intensively fought against), had to be aroused in such a way that the wish to get help through insight was greater than the anxiety of abandoning the neurotic defense system. The therapists had to understand that the latter was necessary as a safeguard, to protect the fragile "body-self" (24) from the unaccustomed drive breakthroughs of puberty.

In our opinion, it proved beneficial to personify an old family pattern with a new variation. The "father" administered the strengthening intravenous infusions, at the same time mobilizing archaic fears of being overwhelmed, and thereby introduced the process of working through. The understanding "mother," who did not demand symbiosis, provided new possibilities for identification. It was probably also important that we saw the patients once—at the most, twice—a week, therefore never letting the relationship become too "dependent."

Miller's (43) reference to the "narcissistic neediness" of the mothers of her patients led us to the important role of the grandmothers in this "family drama" (51). It is interesting that Miller, who worked not with anorectic patients but with narcissistic-obsessive-compulsive and depressive patients, stated, "When we speak of the narcissistic neuroses, we mean . . . the eternally hungry, who cannot obtain anything to eat or, to choose another image, must throw up their food over and over again." This sentence accurately displays not only the dynamics but also the symptoms of anorexia nervosa.

In their regressive helplessness and with their narcissistic transference needs, these patients know how to skillfully mobilize feelings of omnipotence in their surrounding referential figures. This ability is also probably the reason for the diverse rivaling disputes regarding competency within the clinical sector reported by Thomae (56). In this respect, it seems important that as few therapists as possible be involved. Among others, Tolstrup (57) has also supported this view. We believe this is an advantage of outpatient therapy in the dyadic-relational system.

With the combined internist–psychotherapist procedure, a technique was tried out on an outpatient basis that had already been anticipated by Sigmund Freud (quoted in reference 56) for anorexia nervosa and that was recently picked up again by Thomae (56). Previously, however, it had been advocated primarily on a hospital basis. Frahm's (13) successes using confrontation therapy with the stomach tube have shown that one not only can expect amazing things of these patients but, to some extent, must do so in order to free them from their narcissistic "needs for fusion." In our procedure, it proved to be not only feasible but also meaningful to treat the patients on an "outpatient" basis; that is, they attended to their daily tasks and also continued to be exposed to the usual temptations and failures in order to find, thereby, the correct proportion of openness and of establishing self-limits (52).

The combination of internal directive-confrontative treatment with a psychodynamic therapy seemed at first to be self-contradictory. However, we came to the conclusion that in practice, the two directions can be combined well. Considering the rigid defense system of the patients, the insights gained would, by themselves, certainly not have had such a relatively early effect on the symptoms. A steady increase in weight was realized by our patients after one to a maximum of four months only after we began to perform the internist therapy at the same time. On the other hand, we believe that only the internist's dealing with these patients would not have been as effective, if it had not been combined with the psychodynamic insights.

Leuner's (39) "daydreaming technique," GAI, is excellently suited to the psychodynamic mastery of anxiety. In the often-impressive symbolic language of our patients, we were able to approach the targeted core problems with the technique of symbolic confrontation. In this process, one patient went into a deep regression back to the symbiotic level, in which food intake and communication are still "one," and she graphically conveyed how her separation anxieties, or anxieties concerning loss, had come about.

Recently, we have been attempting to deal with "the inability to perceive their body as their own" (Bruch, 8) by intensively taking note of the bodily processes in GAI.

In a kind of internal symbolic drama, the patients confront themselves with their "psychosomatic" resistances. For example, they carry on a conversation with the "doorman, who does not intend to let the food go through," or with the colon, which does not intend to give it up. One patient had internalized her mother, who demanded symbiosis to such an extent that in GAI she experienced "how the intestinal walls greedily seize the pulpy food." After the GAI, she made the association, "Like my mother, who cannot let go of me either."

We still cannot assess whether the patients have really mastered their difficult life crisis and, along with it, the requirement for a short-term therapy formulated by Stierlin (52a) namely, "transformation of the moment into duration." Nevertheless, we are of the opinion that the "deep engagement centered on the main issue" (Malan, 41) has advantages over long-term psychoanalytic therapy, especially as we are referring to an age group that is particularly responsive to such a procedure (a point also alluded to by Stierlin).

We believe that the outpatient procedure will prove successful more often than previously, especially since, in view of the "immoderation" of anorectic patients, effective "measures" can be very pointedly applied within the easily comprehensible dyadic therapeutic relationship. At the same time, the increasing ego strength can be put to the test better in the daily trials of life than in the "indulgent climate" of the hospital.

6

The Effectiveness of Confrontation in Guided Affective Imagery in the Treatment of Childhood Phobias

INGE SOMMER

In my experience with children, Leuner's procedure of guided affective imagery is especially well suited to the treatment of phobias and their effects. Perhaps because of their acute nature, anxieties demand acute relief. Because of a scarcity of time I like to choose the procedure that I perceive as being quickly effective, making a virtue out of necessity. In so doing, I have noticed that a child's imaginary confrontation with situations or persons that arouse anxiety has a particularly intensive effect and that not infrequently the symptom disappears after a single session. Leuner (30) himself has described this (symbolic) confrontation.

In the following I describe three examples of an extreme short-term therapy with children having phobic symptoms.

Case One

Andreas (10 years, 10 months old), whom I had already been acquainted with for a few months, had come to my office because of "nervous restlessness," and I had counseled his mother several times. Suddenly, Andreas exhibited a pyrophobia. For days, he was not able to sleep and could not leave the apartment or stay alone. He was afraid that a fire would break out and had to be constantly on guard that this did not happen. As soon as he saw the glow of a fire (as, for example, when the gas is ignited or a match is struck), he screamed anxiously, and particularly in the evening, he checked light switches and wall sockets over and over again.

Although Andreas had never practiced GAI with me, from the start, the "picturing" worked very well. After suggesting that he relax, I presented the meadow motif:

His meadow is large and vacant. It has short autumnal grass. Andreas feels lonely and is a bit chilly. I encourage him to look around for a forest, which he then finds. After that, I urge him to gather wood so that he can make a campfire in the meadow. He collects quite a large pile of branches but soon does not know what to do next. I suggest that he think about how a fire is made. He places small dry twigs on top of each other to light them and later to put on more wood. After searching a bit, he finds matches in his pants pocket and becomes afraid. He wants to run away, but I encourage him to stand there, look at the fire, and describe it to me precisely. It blazes higher and higher. Andreas is trembling; his eyelids flutter. Now, I tell him that I am by his side and will help him. He is supposed to find out what it is that he is most afraid of now. He says that he is afraid that the fire will spread and burn everything, including himself. I ask whether he knows how one can prevent that? No, there is already too much fire, water will not help anymore. I suggest that together we dig a trench around the fire, so that it cannot spread further. That happens. Now Andreas is calmer. I ask him what he would like to do now. He wants to put on more twigs, so that it will flame up again. Gradually, he has a lot of fun doing this. He dances a war dance around the fire like an Indian. When it has burned down, we extinguish the remains with water. I ask how he is feeling now: "Just great, a little excited."

This treatment episode took place six years ago. I saw Andreas only sporadically after that. There has been no relapse thus far.

Case Two

Walli (10 years, 10 months old) came because of a school phobia. She had begun high school three months before and had to be taken to school and picked up again daily. In the morning, she felt sick and so dizzy that she thought that she was going to "fall down." The specific anxiety, which she formulated, lay in her fear that she would never reach home again. She did not know exactly why. She had vague ideas that someone might lock her in the school. She connected this idea with a teacher, an older nun, of whom many of the girl students were afraid.

Walli was an aggressive, very inhibited child, who grew up as the youngest of three siblings in a family with an unusually authoritarian structure. She was the only one who always lived with her parents; her siblings grew up partly with relatives and partly in institutions for children.

The treatment required 15 GAI sessions. From these, I would like to select the third and the tenth sessions, in which I provoked a direct confrontation with the object of her phobias.

Third Session: I suggest the theme, "Imagine your school." The patient is in the darkness, she is afraid, nothing but arches and empty corridors and stairways. No one is there besides herself. She runs down the corridors in panicky anxiety and looks into the classrooms. Everything is empty. I urge her to stop and think about what she would like to do. She would like to go outside, but then she would have to pass through the school gate. She does not dare do that, because the old nun is always sitting there. It occurs to her that she could go to the summer house, which is much brighter. Her sister goes to classes there. However, the summer house is locked and deserted. Now, she is in the schoolyard and finds a meadow with flowers there, but snow is falling from the sky. Just as she is about to pick some flowers, she sees a sign, which says, "Picking flowers is forbidden." Walli leaves them alone. I ask what her feelings are. She is enraged because here everything is forbidden. What would she like to do? "Throw snowballs." She throws snowballs at the sign until it is illegible. Suddenly, the meadow is situated outside the schoolyard, and Walli can go home.

Tenth Session: I suggest the theme of the edge of the woods. Her teacher (the old nun) will come out of the forest toward her. Walli finds it difficult to picture this image. She can see the edge of the woods only from a very great distance. I encourage her to approach the teacher and perhaps to speak to her. Walli shows signs of anxiety as the teacher comes nearer. She is afraid that the teacher might do something to her. She wants to run away, but at the same time, she is curious about what the teacher will do and decides to stay there. I suggest (in accordance with the management models of feeding and reconciliation described by Leuner, 31) that she speak to the teacher and invite her to a picnic. Walli does not dare to do that. I ask her whom she would like to have along in order to feel confident enough: "My brothers and sisters." In their presence, it works. The teacher approaches. She makes an angry, unfriendly face and says that Walli should not always be running around the whole school building. Now, Walli invites the teacher to a picnic (following my advice, in accordance with Leuner's [38] management model of feeding and reconciliation), spreads out a blanket on the grass, and puts down her basket filled with fruit and vegetables, as I suggested. The teacher eats a great deal. Walli is amazed at that and is particularly surprised that the teacher eats with her fingers "like an ordinary person" instead of using a knife and fork. She thinks that surely the teacher does not get enough to eat in the cloister. When everything has been eaten, I ask Walli how the teacher looks now. She does not look as angry anymore, rather sweet-sour and somewhat blank and rigid. She stands up and asks Walli if she has already done her homework, which Walli affirms. I ask about her feelings: "Somehow relieved." Now I suggest that in saying good-bye, she tell the teacher how happy she was that the teacher liked the food. Walli modifies this, saying, "Perhaps you would like to come again; then I could bring along something for you to eat again. And in school I do not run around that much at all; I only go to visit my sister in the summer house sometimes, but if it makes you so angry, then I can stop doing it." "I hope so," the teacher says and leaves. Walli is relieved, feels exhausted, and would like to stay lying down awhile because she feels dizzy.

After this last (i.e., the tenth) session, Walli was able to go to school alone for the first time. In the subsequent four years of more informal follow-up counseling, a relapse of the phobia did not occur.

CASE THREE

Tine (10 years, 6 months old), was referred because of several minor thefts that she had committed in school. Moreover, her achievement in school had dropped off sharply in the previous half year. She had great difficulty in concentrating and had started playing truant.

Tine was the younger of two girls. Both parents were employed. She had no girlfriends and had difficulty meeting people. The small articles that she stole in school belonged to girl students who annoyed Tine and whom she basically envied and admired. The teacher greatly magnified these incidents which intensified Tine's isolation from her classmates.

I needed six sessions of GAI to attain a behavioral change and remove the achievement disorder. In the first session, I employed the meadow motif:

Tine sees a beautiful, big, green meadow with two big trees in it, to which a swing is fastened. She likes it, but she feels alone. I comment that perhaps some of the children from her class will come along and that she should be on the lookout. Yes, three girls are coming now (the same girls from whom she had stolen things), the ones with whom she has a good deal of trouble. I ask Tine how she is feeling: "I don't know; actually I would prefer not to see them." She sits on the swing, the others stand below and whisper. Tine cannot understand what they are saying and does not feel well. I ask her if she thinks she could invite the others to swing. She does that, and the girls are delighted. Each one wants to be first. Tine takes over the arranging. It gets lively, and they all have fun. Each one swings as high as she can. Tine goes higher than the others, because the others do not dare to swing up into the treetops. Once Tine has shown the others her skill, they say they have to go home and leave. Tine stays behind and is a little sad, because she would like to have continued.

Second Session: I have the teacher appear in the meadow. Tine would like to run away, but I ask her to approach the teacher slowly, to speak to her, and to invite her to a picnic. The teacher looks indifferent and serious. At Tine's invitation, she sits down by the tree trunk where Tine is and accepts an apple. The teacher's expression changes from one of indifference to gruffness. She says, "You have to study more." At first, Tine does not respond to this, and I advise her to explain to the teacher how it is with her. She tells the teacher that she really does study a great deal, but that in school, it is as if all the studying disappeared into thin air; only her anxiety remains and she gets upset. The teacher says that cannot be true, because Tine has no reason whatsoever for being upset. Then, I ask Tine how she is feeling. She says that she is furious and that the teacher is stupid. I ask if she dares to tell the teacher what she is afraid of and what causes her nervousness. But she does not dare to do that. The teacher says, "Improve and it will all work out." They say goodbye.

Third Session: I have Tine imagine the teacher. Now, she appears in the middle of the class and is passing back the arithmetic homework. Tine receives a "C" and is happy about that. Then, the teacher says to her, "That could be better." Tine feels angry but does not say anything about it. I ask her what she is thinking now, and if she does not want to tell it to the teacher. Then, she says to the teacher that she is happy about the "C" and that it means a lot to her to get a "C," as she has received "Ds" and "Fs." Suddenly the image changes, and Tine is with the teacher in the meadow of the second session again. Tine has her dog with her, and the teacher is anxious. I ask Tine what she feels like doing. First, she wants to hold onto the dog. She cannot ruin her chances with the teacher altogether, but she has the dog growl angrily. The teacher acts as if she does not notice it, but Tine can see her fear. Then she calms the teacher down and tells her, that it is safe to touch the dog, that he obeys precisely and will do only what Tine says. The teacher pets the dog, says that it is a beautiful animal, and then leaves. Tine goes home and shows everyone her schoolwork. They all praise her for the "C," and as a reward, her mother gives her some money. Tine buys herself an ice-cream cone with it.

I treated Tine with a total of six GAI sessions. In the final three sessions, I let Tine choose the themes. In each case, the focus was her longing for more attention from her mother, which in her imagination was also granted to her in the sense of wish fulfillment.

After this, her achievement disorders abated very quickly. Then, Tine was able to gain a better social position in class. She again made contact with her fellow students and since then has not played truant nor stolen anything.

REMARKS

In the three cases presented, I intentionally limited myself to excerpts in order to demonstrate the management models of confrontation and of feeding and reconciliation in GAI, which Leuner (31) introduced. I believe that the contents and how they developed speak for themselves and do not need any interpretation. Apart from the intended therapeutic effect, the diagnostic value of the procedure cannot be overlooked. The imaginings show very quickly how the child experiences situations and persons that are cathected with anxiety. At the same time, the child himself is being confronted with the situation that is creating anxiety.

Of course, confrontation and feeding within GAI are not the only instruments in treating a disturbed child. However, as effective aids for relieving situations of acute anxiety, they have a key function of amazingly deep effectiveness in comparison, for example, with therapeutic dialogues. The children presented here were given further counseling on a less intensive basis: Andreas predominantly through my counseling of his mother, Walli through analytically oriented verbal therapy at infrequent intervals, and Tine in group therapy at a later time. This follow-up counseling was certainly not without significance for stabilizing the success of the GAI treatment in the sense of character adjustment.

COMMENTARY (LEUNER)

The author competently applied the techniques of symbolic confrontation and the principles of feeding and of reconciliation. I described these techniques in detail within the framework of the intermediate level of GAI. As already shown in the relevant publications, the therapeutic effect of these principles is generally intensive or becomes quickly effective, as in the example of the child with the bridge phobia, which disappeared after four sessions (extended follow-up observation), although it had existed for several years (cf. p. 31 ff). Aspects of behavior therapy are clearly evident in the principle of symbolic confrontation. However, how the mechanism develops can be compared to only a limited extent to desensitization, as it is systematically applied in Wolpe's (59) procedure with the construction of a hierarchy. Although imaginary confrontation with the object creating anxiety in a state of relaxation is essential to both procedures, the hierarchy of anxieties established in an unaltered state of consciousness in conformance with noetic viewpoints does not occur in GAI. Rather, encouraged by the therapist to confront a real or a symbolic situation, the patient develops his own way of proceeding, which he directs by himself according to the intensity of his anxieties and the defense mechanisms connected with them. The ambivalence becomes evident in the case of the pyrophobia. In the case of the girl with the bridge phobia, which I described, the possibility of choosing the symbolic and, at the same time, spontaneously adopted solution of desensitization becomes particularly obvious. In contrast to behavioral therapy, we believe we can explain this unusually rapid desensitization in relatively few sessions not only by the fact that the choice of contents occurs in a very individual way in accordance with the patient's unconscious, affectively cathected associations between motifs, that is, concentrating to the greatest extent on the individual needs and their defenses. We also explain this desensitization through our assumption of the concept of symbolism and by the fact that the respective content is more than the real object. Much more broadly established unconscious constellations are signalized in it. The functional unity, which was more precisely described at another point, of imagined contents and of the unconscious emotional constellations behind them also becomes apparent here. The therapeutic intervention of the imagination has an immediate effect on this often deeply rooted constellation, which determines behavior. In his giving and protective stance, the therapist concurrently offers a considerable strengthening of the disturbed child's weak ego and also corrects confining superego formations. In this process, he plays a decisive part, which classical behavior therapy does not reflect at all. However, we emphasize the point that GAI addresses itself to symbolic structures in their extraordinary experiential breadth and makes corrections in them, in contrast to the exclusive imagining of real anxiety cathected objects by desensitization in behavior therapy.

7

Guided Affective Imagery in Groups of Young Drug Users*

Edda Klessmann

*First published in *Praxis der Kinderpsychologie und Kinderpsychiatrie*, 22 (1973) 225.

General Experience Gained through Working with Drug Problems in the Child Guidance Center of a Small City

The structure of an officially established child guidance center at first seems to be unfavorable to treating drug problems: overworked staff, insufficient space, and lack of possibilities for controlling the clientele, to whom appointments and regulations mean little. On the other hand, there are considerable advantages:

The staff of our clinic has previous experience in dealing with adolescents and is therefore acquainted with the problems and arguments typical of that age group. Suitable test material and even games and toys (table soccer and even the sandbox have been well received) are available. The members of the child guidance team do not need to be afraid that the center will be closed if the drug-related work is not very successful. Finally, a kind of "peaceful coexistence" comes into play, since just as the adults in the waiting room have to come to terms with the presence of the drug users who are also waiting, the reverse is also true. Of course, we avoid unnecessary confrontations by arranging appointments and rooms according to the problem.

Under these conditions, since the spring of 1971, our child guidance center has been increasingly sought out by drug users and more frequently by their parents. Initially, that put us in the awkward position of having to provide help to a larger group, about whom we at first had no concrete conception. We were startled to find that once the adolescents had discarded their seemingly satisfied facade in our presence, they assumed, with an attitude of somewhat magical expectation, that we could eliminate their difficulties without their having to stop using drugs. Strictly speaking, a confrontation with reality had to take place as soon and as thoroughly as possible. On the other hand, we could clearly foresee that a too critical and rapid disillusionment would close off our access to their group, which is very sensitive to frustration.

Our experience in dealing with drug-dependent persons soon showed that the initial discussions usually take place in an unproductive vacuum. Pubertal drug users are scarcely stimulated to work through their problems verbally. After a period of working in the group, however, we regularly notice that these often quite sensitive adolescents establish a "social relationship" with the other group members and with us. One day, for example, the speaker was hoarse to the point of voicelessness. After a short time, the entire group was also whispering, and although smoking was allowed, everyone put out their cigarettes spontaneously. An important therapeutic factor seems to be that the adolescents in the group became capable of foregoing their egocentric compensatory gratifications.

Of course, we were soon forced to observe that the fascination with drug experience is very great in this age group. In memory, even "horror trips" fade amazingly quickly in contrast to positively experienced sensations. Therefore, our drug-dependent clients often succumb to renewed temptations, when we leave them on their own too quickly. An artistically gifted 16-year-old, "ex-drug-abuser," for example, who was tortured by anxiety states at night for three months following an LSD trip, no longer felt immune to a relapse half a year later, although he had at first sworn "never again."

In our opinion, one must patiently allow for a longer period of "compensatory oscillating," as I would like to call it, between phases of abstinence and milder relapses. During this period, however, it is imperative that attractive counterarguments become assimilated in order for it to make sense to abandon drug usage. All unnecessary tension should be avoided and, at the same time, a healthy tolerance of frustration should be made a goal. A perfectionist expectation of success on the part of the child guidance team must necessarily lead to disappointments. In this kind of work, such expectations should be reduced from the start. For example, I would consider it a mistake to set up rigid demands for abstinence in our situation, which is so absolutely dependent on voluntariness and openness. First, as a rule, they would not be fulfilled; instead, the old barrier of concealment, of guilt feelings, and of projections or scapegoating would immediately be set up again, rather than the desired independent abstinence growing from conviction.

Special Procedures

Guided Affective Imagery in Groups and (or) in Individual Therapy

Our initial perplexity concerning the problem of therapeutic possibilities disappeared when we started group work with guided affective imagery (Leuner), an approach using mental imagery.

As an introduction, we used Grünholz's (18) modified technique of "catathymic imagination," using heterosuggestion. It became apparent that those adolescents who were more severely dependent on drugs had considerable difficulties relaxing "autogenically" to the point of reaching a catathymic image sequence. Therefore, we first had to use very precise directive motifs and have the group members imagine these ideas concretely (among other things, a large dark-green glass vase proved useful for us), in order to make possible a gradual transition—under the influence of music, which they selected—to freely flowing pictorial images from the subconscious. At the beginning, we preferred the "silent group," in which each individual concentrated on his own "visions." Subsequently, the catathymic images, the possible influence of the music on them, and also unconsciously triggered associations with the motifs seen previously were discussed in the group. Later, we followed a suggestion made by Leuner and had the group communicate with each other during the catathymic imaginings (for example, during a mutual "climb of a mountain") in order to develop a group fantasy. In this way, a lively group dynamic could be developed more quickly.

In order to prevent the group from "skidding," we paid increasing attention to a steady balance between the group members who were relatively stabilized and those adolescents who were more strongly drug-dependent. When individuals brought up subjects concerning themselves that could not be worked through in the group very well, these problems were dealt with in individual therapy sessions (also often with GAI). In this chapter, it would take us too far afield to go into detailed information about the technique of GAI or to report on individual group sessions. Just one theme brought up by our pubertal girls, one that seems to be typical, is highlighted here. In their imagination, the girls frequently pictured themselves, in contrast to their outward attire, in long dreses, in an enchanting, dancelike movement sequence. Later, during in-depth discussions this vision helped to explain their identity crisis and to show them their latent wishes for the female role or to promote maturation through clear identification.

Creative Productions

We have the adolescents paint or draw or even model with clay a lot, and in so doing, we like to pick up on the motifs experienced in GAI. We have also organized a course with a graphic artist, who familiarizes the adolescents with the basic techniques of representation. Thus, attention is paid to encouraging the connection with everyday reality. In the group's artwork, we see a healthy counterbalance of spontaneous, partly quite original and partly stereotypically mannered pictures (see Figures 1a and b). The experience with passively consumed drugs can thus be gradually converted into actively creative production.

Autogenic Training at Home

Since the danger of recidivism is great at the beginning, even though the group members temporarily adopt an attitude of distancing themselves from drugs, we "prescribe" a program of practice with autogenic training for the adolescents to perform at home. Many of them carried it through very intensively. Apparently, they sense that from this training, they receive internal "support." After they have completed practice at the first levels, we placed particular emphasis on forming intentional statements. We and the adolescents draft the contents of the statements together, making them as individual as possible and adapting them to individual needs.

GAI IN GROUPS OF YOUNG DRUG USERS   111

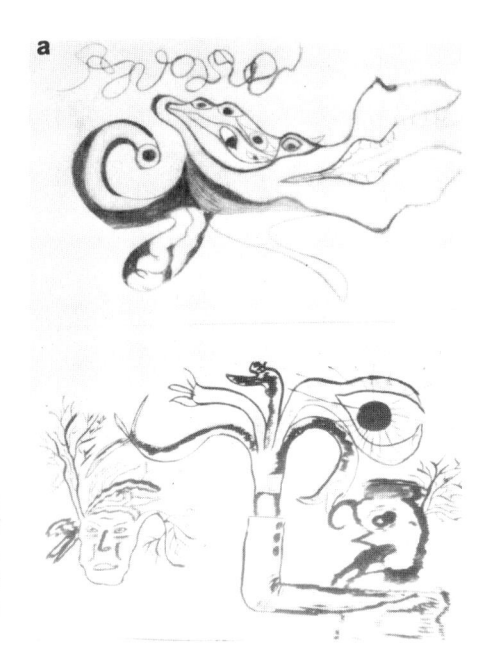

FIGURE 1a: Charcoal drawings by two 16-year-olds as the effects of LSD were abating. The attempt to reproduce the optical visions could not be carried out because the images changed too rapidly.

FIGURE 1b: Charcoal drawing by the same 16-year-old who had drawn the upper LSD picture, here immediately after the GAI session. His image of a "walk through a cave" could be reproduced clearly because it had been experienced much more slowly and definitely than the LSD-trip "experiences."

SPECIAL MEASURES FOR REINTEGRATION

We have our drug-dependent clients take the Hamburg Wechsler Intelligence Test (HAWIE) and the Inclination-Structure-Test (NST) (according to Keller,63) as regularly as possible. In so doing, we are often able to call attention to talents and/or inclinations that have not been previously apparent to the clients. We also encourage them to bring in their own poems, creative productions, and musical performances in order to promote their confidence in their creative abilities. Subsequently, we attempt to make the insights gained from the results useful for activities relating to school, tutorials, profession, and recreation in order to counteract the bent for passivity.

From our test results, which have shown no statistical significance, we will mention one conspicuous point from the HAWIE: Of 16 drug users who were still under the influence of drugs when the tests were taken, half had pronounced coordinational difficulties in the mosaic test (in one case, these difficulties were no longer evident in a control test taken after a long period of abstinence). In contrast, the capacity for memory and concentration (repeating numbers, number symbol test) was not impaired to the extent that we had expected.

Parental Counseling

Whenever possible, we provided in-depth counseling for the parents. Lately, we have preferred open, clarifying discussions with the adolescents and their parents in joint sessions. In such sessions, family conflicts that have existed over long periods of time and are latently smoldering can be dealt with. Occasionally, when separation from the parental home for a long period of time is necessary, a therapeutic group living arrangement is provided. When drug dependent clients are placed in a group with somewhat older, stable adolescents (for example, in a group living situation), we often see favorable developments. In many cases, we encourage the parents to take such a step.

CASE ILLUSTRATION

The development of the (now) seventeen-year-old Birgit is presented as an example demonstrating how our work takes shape concretely.

In the spring of 1971, Birgit came to us through her parents' initiative. She had previously been "sent" to two psychiatrists. Her comment: "One of them only wanted to interrogate me and the other launched the same moral lectures as my parents." As a result Birgit was very negative and distrustful when she had a "psychiatric appointment" for the third time.

These notes, which her mother secretly copied from her diary, serve to document the situation at that time: "Now I have been smoking exactly nine months and nine days. For about three months, I have noticed that I am dependent on it. It is terrible. I am afraid, awfully afraid. Now I am only living from one pipe to the next. About a month ago, I started taking LSD. I cannot get out of this self-made prison anymore. Physically I am still feeling fairly well, but psychically I have reached zero. Someone who has never been in such a situation cannot imagine anything like it at all. I cannot form clear ideas anymore; I no longer know what having willpower is. If K. were not around, I would have already put an end to myself. I gave K. an ultimatum: either K. and I go away together this summer or, if not, it's all over for me. In addition to my miserable situation, there are my parents. Most of all, I am sick and tired of their damned authority and their old-fashioned ideas . . . "

In the first interview, we attempted to analyze the current situation as objectively as possible, for example, the school situation and the calamitous familial situation. We succeeded in arousing her interest in a test relatively soon, "To see what I'm still capable of doing at all." An IQ of 123 on the verbal part of the HAWIE showed a still very good effectiveness. On the other hand, in the practical section, an IQ of 97, with obvious deficiencies in the mosaic test and in placing figures, showed a discrepancy compared to the first part. Even Birgit, who generally seemed relatively indifferent, was struck by this.

In the control taken one and one half years later, this time not under the influence of drugs, the MT-and FL- results were above average for her age. After the test, we were able to enter into an objective discussion "without morals" about drugs and her problematic future. Birgit made no secret of how badly she was feeling but was unable to distance herself from taking drugs. She refused to be placed in a hospital. Instead, she brought along her boyfriend, who was equally dependent on drugs, to her next appointment. Both were considering whether and to where they would "take off" in order to begin a "new life."

We succeeded in pointing out the illusion of such an undertaking and in inducing Birgit to spend 14 days with relatives, whom she valued a lot. For the first time, this distancing brought with it the hope for a genuine new beginning. However, in the period that followed, stresses (separation from her boyfriend, failure in school, arguments with her parents) led to relapses over and over again. In time, however, these became shorter and settled down to "weekend trips." She found these to be very pleasant. Afterward, however, she was always particularly "depressed," as she admitted herself. After half a year, she took part in a treatment group with GAI and practiced autogenic training at home. Now, she also painted and sketched a lot, though, of course, at home rather than in the group. However, she brought her creations along to her individual therapy sessions, and we were increasingly successful in dealing with early childhood conflicts using an "uncovering therapy." Now, for the first time, Birgit admitted that three-quarters of a year previously she had been far enough along to "shoot up." But the "guy" had let her down, which had made her "incredibly furious" at first. However, now she was glad that it had not come to that. (The "shooting up" was to have taken place at a time when she had just begun to gain a foothold with us but did not yet feel secure enough to reveal her deeper problems to us.) Now she also reported about dreams in which she gave herself the needle or received a needle in her vein from a male person. On one occasion, the needle was gigantically big and filled with a yellow fluid (sexual symbolism). She experienced the dreams as fascinating and frightening at the same time. In GAI with the group, she frequently had sadomasochistic visions, in which she saw herself being run over or imagined her boyfriend's death. As this imagery disturbed her a great deal, the theme was taken up in the individual therapy sessions.

After a year, she was drug-free (no relapses for a year, either). During the past summer, she passed her examinations, whereas previously, being expelled from school had seemed scarcely avoidable. A half year before this, she had made up for the gaps in her studies through intensive tutoring. In the meantime, she is in practical professional training and seems outwardly stabilized. Her attitude toward experiences with drugs was, however, more realistic during the period when she felt bad than it is today. In the meantime, she can scarcely remember the negative things from the time of her truancy and "being flipped out." She destroyed her diary, as a possible reflector of reality. She recalls the fascination of the drug highs from that time much more intensely. If she can still keep a distance from renewed drug use, it will be because she experiences a confrontation with reality over and over again in therapeutic discussions and has become more conscious of responsibility through a stance of solidarity with the child guidance center. The latter is probably true partly because we had pleaded her case in criminal proceedings.

Numerical Data (Interim Results in the Spring of 1973)

Of the 109 drug users known to us (occasional users are not included here), 68 came to the child guidance center frequently. Another 18 came to us only 3–10 times; 35 were in individual treatment, and we also treated 35 in groups. At the time of the first interview, 24 users were in exceptionally poor condition (not only as far as their health was concerned but also with regard to their social situation).

They were mostly multiple drug users, using predominantly cannabis but all other pertinent drugs as well, including alcohol.

At present, we still rank about half of our pubertal clients, not dependent on drugs, among the "compensating vacillators," who now and then "smoke a joint" or even take LSD. However, all of them have changed their attitude toward drug use in the direction of greater distance. At the moment, not one case can be regarded as a "dropout" as far as health and social situation are concerned.

Of the 11 drug abusers known to us, 4 (with respective court cases) apparently found their way out of a "dropout phase" easier after they had found support not only in a group with us but also outside in partnership or group ties. Only 1 adolescent began shooting up and disappeared in the dropout scene after he had been with us (four individual sessions). In the meantime, we have also learned from 10 psychically strongly dependent multiple drug users that before they came to us, they, too, had been "just short of shooting up" (some having already dreamed about it, as did Birgit). However, none of them did actually shoot up, although, as we learned, reaching for the needle can exert a frightening power of attraction on adolescents as a rite of initiation (see Figure 2).

We discovered therapeutic possibilities precisely for this age group in the treatment offered to drug-dependent pubertal adolescents at the child guidance center of a small city in Westphalia. Besides the personal commitment of the team, these possibilities required relatively few resources other than those usually available to the child guidance center.

Our psychagogic and psychotherapeutic interventions corresponded in many respects to those which we also used with other clients of this age group. Some can probably also be used with older drug abusers. That is particularly true of musical GAI (Leuner, 38) in groups, which is frequently experienced as a welcome substitute for the "consciousness-expanding drug" and automatically leads to a psychodynamically oriented therapy. The question arises whether our patients were particularly favorably selected and would have found their way in any case. We cannot prove the opposite. Nevertheless, we believe that stabilizing someone who has "flipped out" becomes easier if he is no longer constantly in the midst of the negative feedback from the drug scene.

Therefore, with our stabilizing counterweight, we attempt to ini-

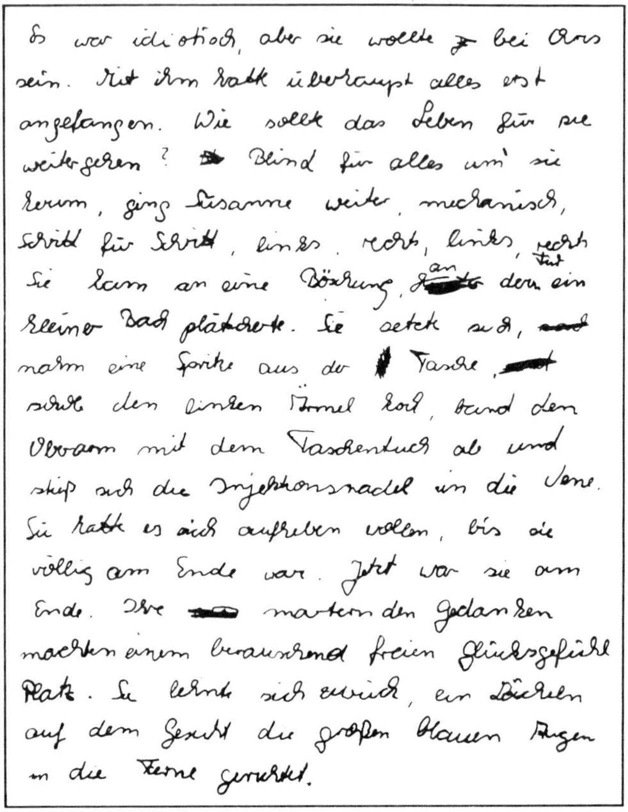
Es war idiotisch, aber sie wollte bei Chris
sein. Mit ihm hatte überhaupt alles erst
angefangen. Wie sollte das Leben für sie
weitergehen? Blind für alles um sie
herum, ging Susanne weiter, mechanisch,
Schritt für Schritt, links, rechts, links, rechts
Sie kam an eine Böschung, an den ein
kleiner Bach plätscherte. Sie setzte sich,
nahm eine Spritze aus der Tasche,
schob den linken Ärmel hoch, band den
Oberarm mit dem Taschentuch ab und
stieß sich die Injektionsnadel in die Vene.
Sie hatte es sich aufheben wollen, bis sie
völlig am Ende war. Jetzt war sie am
Ende. Ihre marternden Gedanken
machten einem berauschend freien Glücksgefühl
Platz. Sie lehnte sich zurück, ein Lächeln
auf dem Gesicht die großen blauen Augen
in die Ferne gerichtet.

FIGURE 2: From the autobiographical "novel" of a 15-year-old girl who had experimented with hashish and was acquainted with a narcotic user. We frequently find a fascination with the anticipation of the orgasmic ecstasy of the "hit" at this age. "It was idiotic, but she wanted to be with Chris. It had all started with him. How was life supposed to go on for her? Blind to everything around her, Susanne went on, mechanically, step by step, left, right, left, right. She came to a bank, at the foot of which a small brook babbled. She sat down, took a syringe out of her pocket, pushed up her left sleeve, tied a handkerchief around her upper arm, and jabbed the injection needle into her vein. She had wanted to hold it up, until it was all gone. Now she was gone. Her tormenting thoughts were replaced by a dazzling free feeling of happiness. She leaned back, a smile on her face, her big blue eyes looking into the distance."

tiate a reconciliation between the adolescents, whose strivings for independence during puberty are full of crises, and their earlier circle of relationships. The final therapeutic goal, the total break from drug use, is often difficult to realize, particularly with hard-drug users, as is well known. This seeming futility is probably one of the primary factors keeping many potential helpers from becoming more intensely involved in work with drug users. Our previous experience may make it clear that, at least, one does not have to capitulate before the equally important and rewarding short-term goal of keeping the adolescent soft-drug users from the impending use of hard drugs.

Today, after the hippie wave has ebbed, "consciousness altering" is no longer the preoccupation of young people who fail at the compulsions of achievement and who consider themselves among the elite. In 1976, the drug users come to the child guidance center, if they come at all, with other expressed motives. They complain primarily about difficulties relating to work and to social contacts.

One might think that our musical group GAI of that time, which all but met the needs for "consciousness expansion," would elicit less interest today. However, we have found that the same therapy gets across in the same way. Currently, it is offered as a means of facilitating intensive self-awareness in the group and as "communication therapy."

Here, I must include the fact that at present, we no longer counsel any "purely drug groups" but "mix" them with other neurotically disturbed adolescents (for example, with previously treated anorectic patients). Among these drug users are also some who already have a criminal record or have been shooting up for some time, but who are nevertheless motivated and are able to come on an outpatient basis.

We have tried to illuminate the problems by using role playing in such a mixed group. However, it became apparent that most of these clients could not yet deal with this direct confrontation. Then, in the musical GAI, we found that, for example, a participant in the "psychodrama group" who was incapable of saying even one word in that group spontaneously brought up her problem in the very first musical GAI session. She had felt so terribly isolated. In fact, she had, unconsciously, clearly set herself "apart" and understood, as did the group, that this was no coincidence.

It would take us too far afield to go into detail about specific features of musical GAI here, which have apparently made it possible for severely disturbed patients to enter into the group experience. Additionally, it should be mentioned that in any event, the therapy we used has not lost its applicability to today's different "drug generation" (provided that an openness to therapy exists).

8

Guided Affective Imagery in the Short-Term Psychotherapy of a Drug Abuser*

HANS-MARTIN WÄCHTER AND HANSCARL LEUNER

*First published in *Praxis der Kinderpsychologie und Kinderpsychiatrie*, 23 (1974) 81.

Reports have repeatedly been given of the significance of the mental image technique of guided affective imagery (GAI) in the psychotherapy of drug-dependent persons. Previously, these reports have always involved the use of mental images in a group situation aided by music.

The present case report deals with the individual therapy of an 18-year-old adolescent high-school student. He was the patient in an investigatory series with GAI, in which 14 randomly selected patients were subjected to treatment in short-term therapy, which was limited from the start to 15 sessions over an eight-week period (2 sessions per week). With regard to the technical problems of short-term psychotherapy in general, one can refer to Leuner (36) and in the psychoanalytic field to Malan (41) as well as to Beck (5).

In addition to the clinical data, we used a test battery, which we put together and which Pudel (64) worked out statistically, as well as semistandardized interviews to establish the results. A detailed presentation of this investigatory schema and the problems connected with it will be forthcoming at another time. The test battery consists of the Giessen list of complaints (67), a list of psychopathological complaints that we drew up; the questionnaire test by Brengelmann and Brengelmann (61) with scores for extra- and introversion, neuroticism, and rigidity; and tests for manifest anxiety by Taylor (66) and the Minnesota Multiphasic Personality Inventory (MMPI). The orienting examination of intelligence was done with the Raven (65).

Guided affective imagery was introduced to psychotherapy as a psychodynamically based approach employing mental imagery, and its therapeutic significance has been documented in methodological and case studies (Leuner, 38).

The similarity between experiencing in GAI and the experiential categories during a drug high has been pointed out. In both cases, an altered state of consciousness is present, and the subject perceives optical phenomena in both cases, so that based on his experiences during "highs," he is on familiar ground experientially when GAI is used in his psychotherapy. An imperceptible transition from the drug experience to the therapeutic experience, so to speak, becomes possible. For this reason Klessmann (22) went so far as to consider GAI the "method of choice" in treating this group of patients of drug abusers.*

The following transcript of the treatment, which in part is reproduced verbatim, expresses the peculiarity of the experiential categories mentioned and the facility with which the abnormal neurotic attitudes and emotional deficiencies of the drug-dependent person become evident. The unconscious motives (for instance, the lack of security, the longing for tenderness and understanding) are spontaneously expressed here without the necessity for detailed interpretations. It is further important to note how intensely and vividly the patient recapitulates a drug experience in GAI, almost as a piece of "need satisfaction" in the sense of a compensatory satisfaction without using the drug. From this recapitulation, the patient began to see that GAI was a surrogate for drugs. Recognizing this was what finally led him to give up drug abuse. Through this process, the client stopped using drugs without even the slightest "pressuring" or any other kind of intervention by the therapist, but solely because of the developing transference and the insight he had gained through therapy. He then picked up the regular high-school course of studies again.

Finally, we would like to emphasize the point that the treatment technique used was within the range of the elementary level of GAI (predominantly training procedure) (40), that is, at the technical level of a therapist (W) still in training for the title of "psychotherapist." He

---

*See Chapter 7.

had learned GAI in courses and in personal guidance (L). The treatment was carried out under L's supervision and was, from the start, limited to 15 sessions. The postobservation period lasted for 10 months.

The 18-year-old high-school student Julius F. came to our office in July 1972 accompanied by a hard-drug user who had begun psychotherapy with us on his own initiative. Julius complained of depressive mood states, dejectedness, fatigue, and sleep disorders. In addition, he had major work-related problems in concentration, which had caused considerable difficulties in school. Even after repeating the eleventh grade, he had not achieved the passing level for his class in two major subjects and finally had to leave the school. After that, he was "fed up" and wanted to suspend further school studies and work or hitchhike through Europe.

After having taken drugs for approximately two years (first rosimon, then hashish, and finally LSD), he became one of the best-known authorities of the drug scene in his hometown, a small city in Lower Saxony.

In addition, Julius showed considerable lability in his self-esteem and had contact problems in general, which were even more severe in the company of girls. He felt partly drawn to them, partly repelled by them and was afraid of being homosexual.

His symptoms (school problems, depressive moods, and social withdrawal) began in August–September 1969. They grew worse at the end of 1971. In October 1971, Julius attempted suicide using rosimon and again in April 1972 using limbatril (after a horror trip). The situation triggering the later attempt was the change of residence of the girlfriend who had introduced him to the "scene." During the later suicide attempt, Julius was brought to the local hospital and was subsequently referred to us as an outpatient. Our first psychotherapeutic attempt consisted of two sessions with psycholytic treatment (using the hallucinogenic drug psilocybin). Soon, however, Julius stopped coming, and the treatment could not be continued.

Previous History: Birth by Caesarean section, was not breast-fed, delayed motoric development (walked at 1½ to 2 years of age), no phase of defiance, was reportedly a "very good child." Primordial symptoms: thumbsucking and screaming during the night until the age of approximately 7 or 8 years, nail biting up to the present time.

After their marriage fell apart, the parents divorced when Julius was 3 years old. He grew up with his mother, who seemed to be very neurotic and had been under a neurologist's treatment for a long time. She attempted to control how her son lived his life, suppressed any tendencies toward independence, prohibited him from having contact with friends and girls, and also rejected all separations from the family home. The patient: "She treats me like a child; I am her child and she doesn't want to let me go." She tried to influence him using various means: authoritatively, by moral extortion (she simulated suicide attempts), and also by means of the "friendly route" along the lines of "I give you something, you give me something, too" (brought him breakfast in bed, put money for cigarettes on his bedside table, etc.).

Julius saw his father, who lived in a town several kilometers away, as a "negative authority." Julius had no emotional relationship with him, thought he was "stupid," could not converse with him. However, Julius was afraid that his father would also put claims of possession on him. Each parent reproached him for siding with the other parent. Julius had a good relationship with his stepsister who was nine years older (mother's premarital child). However, after the sister got married, the relationship had become more distant.

Other referential figures: Julius had considerable contact problems. He could make acquaintances only along the lines of "Got any dope?" His great need for support and tenderness from others was reflected in his ambivalent friendship with a student who was two years older than he. This friend pushed him into the "daddy–little son" role. His friendships were mostly structured in a triangle or quadrangle, in which a girl played the main part, along with one or two male friends. In his last relationship of this kind, "she" had played a "positive mommy-role" for him.

Sexual Development: Julius has masturbated since the age of 13 and has "enormous" guilt feelings about it. He thought his body was "filthy and disgusting" and that that was the product of his mother's upbringing. Whenever he was alone with a girl, he was afraid of sexual approaches, became passive and was repelled by active or demanding girls, and was afraid of failing.

At the first meeting with the therapist, the red-headed, outwardly somewhat unsightly and simply (almost sloppily) dressed Julius made the impression of being contact-inhibited, depressive, and grouchy. In addition to his marked acne vulgaris, his disconnected, peculiar way of speaking was conspicuous. He used a great number of expressions from the drug scene as well as frequent gestures and grimaces or unarticulated sounds (as, for example, "pfft") to replace words that he could not recall. He was quite intelligent (Raven III*) but seemed conspicuously inarticulate for a high school student. Passive tendencies along with considerable psychic pain were unmistakable.

Diagnosis: Depressive mood state, work-related concentrative disorder (school difficulties), and drug abuse with an affectively labile, insecure personality in a maturational crisis, broken home (attachment to mother, separation problems, guilt feelings relating to masturbation).

Development within GAI: As is within the range of a short-term psychotherapy, the treatment had been limited to 15 sessions (of approximately 55 minutes each) from the start with the patient's knowledge. It was performed on an outpatient basis. Recording the semistandarized anamnesis required four sessions. The patient took the psychodiagnostic tests (all questionnaires)

---

*Level III corresponds to above-average intelligence (65).

on his own with supervision; altogether they lasted approximately three hours. The GAI exercises lasted as a rule from 35 to 40 minutes; the rest of the time was devoted to pre- and postdiscussions with the goal of illuminating the current problems with his actual situation. Interpreting the contents of GAI was avoided on principle or was limited to brief hints, when associations struck the patient himself, when—as he said—the "consciousness-expanding lightbulbs" went on.

Even during the taking of the psychodynamic anamnesis, Julius developed a positive transference to the therapist, who offered him tolerance and understanding. The use of drugs (hashish, LSD) in general and Julius's experience with them were discussed objectively. The therapist did not condemn drugs, made no value judgments whatsoever, but let it be understood that he did not consider the use of drugs necessary and that Julius could probably get his problems under control in another way, too.

In treatment, Julius's diverse problems were expressed, in part, symbolically in GAI and, in part, directly in discussions. At first, this concerned particularly the drug problems, which dominated the first three sessions. Thus, the main motivation for using drugs also became indirectly clear, namely, the longing for security, the need for supportiveness, and the search for human contact. Even in the first session, Julius was able to follow the GAI procedures very well, and he could instate several standard motifs of the catathymic panorama. He found himself in a meadow with rich, dark green grass at the edge of a small forest and then came to a wildly romantic valley. The seclusion of the valley and the rocky slopes were familiar to him (image from memory) and gave him a feeling of security. He then found a crystal-clear brook, which he could pursue to its source, but the water was cold. There were only stones and no vegetation in the vicinity of the source. After some hesitation, he climbed a mountain and observed the panorama. He looked down on a wooded mountain landscape. To the west, he saw some fields and villages. This view seemed boring to him. Having returned to the first valley, he felt lonely. He lit a campfire. Then, several acquaintances, girls and boys from his hometown, appeared.

In the second GAI session, his need for support and the drug problems became particularly obvious. Julius first found himself in a meadow. He passed by a tall, dark green tree, a blue spruce with piercing needles. Nearby stood a big, old, gloomy, almost unreal house. Passing through a big door and past several gloomy, cold, and depressingly furnished rooms, he reached the top floor. He found his friends assembled in an unpretentiously furnished and colorfully painted room. They were sitting around apathetically and sadly, each one preoccupied with himself, until someone made the suggestion to take "acid." Julius did that, too. He then relived an LSD trip very intensely in GAI.

He described, "A meeting is taking place. S. (a girl) is sitting on the rug and drinking vodka. Z., her boyfriend, is sitting on a couch with a joint in his mouth. A. (a friend) is sitting in the corner and looks at me sadly with his big black eyes. He is probably in his depressive psychohorror again. B. (a girlfriend) is also sitting on the rug and is on an intellectual trip at present. Perhaps she considers this whole scene highly dramatic. Then I. (his fatherly friend) arrives. Since he represents the "acid daddy" for me in any case, I lose my fear and take a whole tab with him. . . . The "acid" takes effect. I. looks like a skull . . . then his face changes. He seems calm and trustworthy. I would like to lean against him but nevertheless do not do it. I sense the same thing when I look at S. (his girlfriend) but not as strongly. Then everybody makes me sick. Perhaps because I see everything so extremely and know that I am exactly the same as they are. Music is playing in the room. Sad, it is a little bit uncanny. It is about angels, Satan, and the spirits of the dead. Once, I listened to it all night long with I. Perhaps one passage from the lyrics expresses it more clearly:

> . . . The sun is shining in the night. Zombies walking through the broad daylight. Zombie you done lost your soul. Zombie you don't never get old. Lord of darkness, king of light, come, come, come here on this stormy night.

"I run out of the room, storm through all the rooms, and am looking for someone. Perhaps I., but he is upstairs. S. (the girlfriend) wants to go home; I. (the friend) drives her. They ask me whether I want to go along or not. I want to be with I., but I stay. What difference does it make, I think . . . " (from the transcript of the second session). Finally, the pseudohigh subsided under the guidance of the therapist. On returning to the initial meadow, the tree had changed. Its branches were spread so that Julius could touch the trunk. But it was rough and hard, and he could not lean against it.

Parallel to the experience in this session, Julius's general condition, including his depression, worsened. He had a bad weekend.

In the third treatment hour, the drug scene also played an important role, whereby the lack of contact and unconnectedness of the patient was expressed even more intensely. He wandered aimlessly through his hometown, visited a familiar bar, but felt alone everywhere. He refused offers of the imagined persons to take LSD and sat around helplessly. Then, he went for a walk through the city park with a bottle of red wine and a girlfriend from school in order to provoke people. Finally, he went back to the bar, where he met his friend, who played the older brother and also sometimes the father for him. Then, both began to smoke hashish. In so doing, the patient relived a scene that had in fact happened a half year previously, but now this experience was much more intense. He found himself in a forest and sensed an unlimited feeling of happiness, when first this friend took him into his arms and then a girl embraced him; he felt secure and loved.

The wish for security and the drug problems appeared together again in the seventh session; however, the former was in the forefront. In his imagination, Julius flew to Africa by plane and suddenly found himself in a beautiful and wild jungle. In a clearing in the woods, there stood a native village, whose inhabitants invited him to have some tea and smoke marijuana. They were merry, unconstrained, and natural. He understood them although they spoke a foreign language. In the evening, he lay down to sleep in a hut with a girl and a man on either side and felt cozy and secure.

Presumably under these impressions, the patient took LSD again two days later. However, he had guilt feelings toward the therapist, and feelings of anxiety and nausea, and resolved that this was the last time. A week before that, he had taken half a tab in a moment of loneliness and depression, had not felt anything particular from it, and discovered to his disappointment that LSD did not give him what he was looking for. GAI had in the meantime become a substitute for him.

In the middle of the treatment series, he compared the effect of GAI with that of psycholytic therapy, with which he had become acquainted (two sessions). He preferred GAI to it, however, because he was usually calm and more well balanced afterward. From this point on, the drug problems disappeared from the contents of GAI. The wish for security continued to appear, also within other motifs. Several leitmotifs passed through his imaginings now and overlapped each other.

In the fourth session, Julius fulfilled his wish for warmth in a peculiar way. First, he imagined an endlessly broad meadow after he had followed the course of a river, which flowed into a lake. Between the endless area of the lake and the large meadow, he felt lonely and was freezing. There were frogs and toads hopping around in the meadow, which filled him with great disgust. Also in the water, there was a slimy, brown monster covered with scales moving around, partly resembling a crocodile and partly a snake. Finally, he found a house, which was actually an empty stall, and with the straw that was lying around, he made a fire to warm himself. The house went up in flames immediately. Rats left the burning house and chased away the frogs, whereas the patient lay down in the flames but did not burn. Instead, he was cozily embraced and made warm by the flames. Then, he fell abysmally into a burning city, and then, after having left this city, he found a girl in a meadow who resembled the exotic-looking protest singer Joan Baez. With her singing and guitar playing, she imparted to him a pleasing and calming feeling. She appeared twice more in later images and conveyed calmness, security, and also "eroticism" to him.

In GAI Julius finally perceived a set of ideal parents. While walking along the beach he came upon a fishing village. In front of one of the houses a married couple, fishermen, both blond and with striking facial features, spoke to him and invited him to tea. In the fisherman's house he felt perfectly at ease, secure, and satisfied. After this experience he felt considerably better, his mood changed to the positive and now he could also sleep better.

After this session the motif of the dangerous mother gained in significance in various forms. On a sandy beach he was threatened by a gigantic, disgusting, and dangerous sea spider (the symbol for his possessive mother). His description follows (transcript of the sixth session):

"I am on a sandy beach. In front of me is the ocean with its foaming surf and behind me are the dunes. The surf throws a sea spider onto the beach. It is disgustingly slimy and has a brownish color. Its legs are hairy. The creature is crawling toward me. I am disgustingly afraid of the spider. I feed it fish (at the therapist's suggestion). It grows even bigger and greedier. Then, it does not want any more fish. I throw raw meat in its direction. It swallows the meat greedily. It has goggle eyes and a very large mouth. It gets bigger and bigger, until it is almost as large as I am. I cannot get rid of the feeling that it wants to eat me up. After I throw it a few more hunks of meat, it trots back into the ocean. I continue walking along the beach."

In the following sessions the patient was frequently threatened by sea spiders in the water or by black, lurking spiders in the cellars of houses that he entered.

On another occasion (tenth session), he walked past a cow pasture in a meadow with an idyllic atmosphere, where several cows "stared stupidly" at him. Fortunately, they were separated from him by a barbed-wire fence. He wanted his peace and quiet and went to a bar to play cards.

Finally, in the fourteenth session, in addition to various sexual symbols, an insatiable octopus appeared in the sea, which threatened him and greedily sucked up the fish that the patient threw to it (therapist's suggestion). The octopus grasped him by the arm and threatened to swallow him. At the last moment, Julius was able to repel him with a harpoon. Then, the octopus clasped the harpoon, snatched it away from him, and swam away. The ambivalence toward the mother was expressed in various additional imaginings, also with regard to his relationship with girls. Frequently in GAI, the girls' breasts began to grow and took on threatening shapes. He was repulsed by them and filled with disgust.

In several GAI sessions Julius dealt with his relationship to the female sex and to sexuality. Initially, girls he knew from his circle of friends appeared, who played with him and occasionally took him tenderly into their arms, all this, however, within the group of acquaintances. Later (eighth session), he found himself in a meadow under a plum tree and had the feeling he was doing something forbidden as he picked the sweet fruit. He fed the plums to a swan swimming on a lake, that greedily bit open the plums, and ate only the pit. Then, a slim blond woman appeared; now, the swan let itself be fed by her. The woman invited him to her place, but he found her to be disquietingly superior, cold, and arrogant and felt repelled by her. In the following (ninth) session, which he described as one of the most impressive, he was with a beautiful young girl in a valley that he knew. He had arrived there by passing through a large iron gate. He embraced the girl; they kissed each other and lay down in the grass. The difference between this girl's delicate, soft kiss and the kiss of his "fatherly" friend struck him. He perceived it as very nice to be able to be together with the girl and was glad that he had time to let the relationship develop slowly. In a later session (twelfth session), he met another blond girl at the same spot, who romped around the valley wildly and sensuously with him. Here, he suddenly had the feeling that he was totally hemmed in and could not move anymore. Later, womanhood appeared to him as a golden statue of a goddess in a South American temple. He sat down at the feet of the goddess, who looked down on him good-naturedly, and drank some of the blood-red wine, which was in a silver pitcher next to the statue.

In the course of GAI, the motif of the house went through an interesting transformation. First, it appeared as a gloomy building or an old castle with long corridors, with dark and coldly furnished, depressing rooms and a formidable, dark cellar with cobwebs and black, lurking spiders. It then changed into a burning stable. Here, Julius lay down in the flames. The house in which he lived with his mother appeared toward the end of treatment. He put a bouquet of flowers on the roof, whereupon lightning struck, and flames blazed up and destroyed the whole building. In the night after this session, the patient saw the rough construction of a new house in his dream. He interpreted this spontaneously as a new beginning in his life.

Results After Termination of Treatment (15 Sessions)

1. According to the patient's *subjective perception*: His mood has improved, now he suffers only sporadically from depression and has become more self-confident and more frequently "cool." He no longer speaks of fears of being homosexual and reports that he is beginning to have a lively interest in girls. He also believes that he has become more self-critical; his desire for LSD is gone. In the meantime, he had once taken one more tab, but it had been a "bummer" for him and he wants to give it up completely. He has become more sensitive and open to the stimuli in his environment.

2. *Test results*: On the Brengelmann and Taylor scales, after treatment was completed, there was a rise in the value of rigidity from 8 to 12 (which can be interpreted in the sense of ego strengthening) and a drop in the anxiety level from 35 to 25 points (circa 25%). There was no influence on intro-extraversion and neurotic tendency.

The psychosomatic complaints (Giessener list of complaints) decreased from 52 to 42 negative points; the psychic symptoms of the Göttinger list declined from 43 to 24 negative points.

A comparison of MMPI test data before and after treatment showed a distinct improvement in the patient in the sense of a drop in the scales characterizing anxiety, insecurity, irritability, tenseness, and worry concerning bodily functions, as well as social introversion, and a slight rise in the scales suggesting an increase in activity. The strong feminine attitude relative to distribution of interests was likewise clearly reduced (Mf: 79–69).

3. *Social changes*: Julius had left school before beginning treatment, because he had not passed two subjects, and had toyed with the idea of giving up high school completely and working for one or two years or hitchhiking through Europe. He decided, on completing treatment (October 16, 1972), to finish his school education at a boarding school. From what he said, the treatment enabled him to make this decision, which also signified separation from his home and mother, and to carry it through.

Two months after terminating treatment (December 27, 1972), Julius visited us again, after he had been accepted in the out-of-town boarding school. He gave the impression of being more mature, more well balanced, and more self-critical and was determined to complete school. To be able to work, he wanted to give up using drugs completely, at least during school. During the vacation, he had occasionally smoked hashish on the weekends. However, he had been able to distance himself from his hometown clique and also viewed his relationship to his mother more soberly. Only occasionally did he become depressed for a short time, particularly on weekends, because then it was boring at the boarding school. He had relatively good contact with his fellow students, although in his opinion relationships in this boarding school were quite precarious, as the system was authoritarian and he felt himself being "pushed aside." Thus, for example, he missed having a referential group but wanted to see it through anyway and complete his high school diploma there. He had become a member of a study group for political education and took an active part in their events.

In his repeated visits, the last in April 1973, an increasing stability of his condition emerged in spite of occasional reactive vacillations in mood on weekends and during visits at his mother's home. Only during vacation did he smoke hashish with his old clique again. His interest in girls had increased, and he talked about his contact problems. He was happy about his good school grades.

The discussion of the case can be kept brief. We initially went into detail about the possibilities of GAI in the treatment of drug-dependent persons. The fact that the therapist kept largely to the technical guidelines of the elementary level of GAI indicates the relative simplicity of applying the procedure in therapy.* At the same time, its effectiveness in quantitative and qualitative respects becomes clear both in the imaginal contents manifested with the release of the problems and in the immediate results that were attained. For the present, the improvement can at least be characterized as "good," if one uses Boehm's (60) categories for the outcome of psychotherapy. However, a more conclusive judgment can be made only after long-term follow-up.

Everyone who is experienced in the field of drug abuse knows that very often the symptom of drug dependency does not enter into treatment; instead, what enters into treatment is the neurotic character structure of the client, which in this case was shown in a multiplicity of character disorders, including the tendency toward depressive moods. The contact problem and the problem of submission became obvious. Apparently, however, the therapy enabled the patient to make a new social beginning. From the more relaxed character structure connected with this beginning, social correctives (such as participation in a political study group, the adaptation to the boarding school situation, etc.) seemed possible. We would like to assess as particularly favorable the patient's own motivation to continue his school education and to move to the boarding school, which involved the mastery of various initial anxieties and frustrations.

Of course, we do not doubt that at least a part of the therapeutic effect was due to the transference, which definitely can be characterized as "positive." Here, however, the therapist stepped into the gap left by the yearned-for father image. The resulting "attachment" also appeared in later repeated wishes to visit the therapist in difficult situations in spite of the great distance from his residence. However, we are still of the opinion that in addition to the new clinical adaptation, the statistically significant improvement in a series of the pathological scores also indicates at least a partially completed and obvious per-

---

*During spontaneous productions, however, an associative procedure (intermediate level) was also used.

sonality change. Finally, it seems noteworthy that such a serious symptom as drug abuse was relinquished, a symptom that one cannot rate very well prognostically because of the compensatory satisfaction connected with it. Within the scope of the whole group of 14 clients given "short-term" treatment in individual sessions with GAI, Julius classifies as significantly improved.

9

Guided Affective Imagery in the Therapy of a Severely Disturbed Adolescent

GÜNTHER HORN

During the years 1966–1968, I treated Ralf using GAI. The development of the therapy, which took a total of 35 GAI sessions and several dialogues, was in many respects so impressive and exemplary for the treatment of a severely disturbed adolescent that excerpts will be reproduced here. The focus of this presentation is on examples of some of the patient's imaginings, which give a vivid impression of the development of the treatment.

The contents of the therapeutic dialogues, in addition to the GAI sessions, cannot be presented in detail here because of the brevity of the presentation. Theoretical reflections likewise play a subordinate role in this presentation.

Ralf was 16 years old when his parents made a treatment appointment for him. The diagnosis read: "Anxiety neurosis, massive father problems, depressive structure." The symptoms were persecutive anxieties after dark, failure in school and absolute refusal to attend school despite above-average intelligence, general disobedience, suicidal tendencies from the age of 11 on, gnashing his teeth at night, frequent headaches, and suddenly occurring diarrhea. Ralf came from a family that was economically well off. His father, who was 65 years old and not very tall, held a leading position. In his bearing, he gave the impression of being extremely matter-of-fact and detached. The father stated that he had lived a very isolated life since his childhood. In contrast, his wife was quite gregarious. She had built "a bridge to other people" for her husband. The father had compensated for his own inadequacy with a successful professional career. During the Third Reich, he had held a high position in the economy. When this became known at the beginning of the 1960s, he was exposed to violent attacks in the local press. Nevertheless, he was able to maintain his position in an industrial firm.

The father had envisaged that his children would be successful in life, too. However, whereas the two daughters—10 and 12 years older than the patient—had been able to identify more with the ideal of the mother and had been moderately successful professionally, Ralf and his brother, who was 8 years older, had failed completely. At the time the patient was referred to us, his brother had already been in psychotherapeutic treatment for two years without having been able to persevere at any particular profession. The only hobby that both brothers kept up intensively was playing the drums. This seemed, among other things, to be a late consequence of an identification with their (maternal) grandfather, who had passed away five years before and with whom Ralf, in particular, had had a positive relationship. The grandfather had been a drummer with a well-known military band. At the same time, in the choice of the instrument of drums, a protest against the "bourgeois" behavior of the parents was probably being expressed, as the parents regularly played classical music. The drums were the only instrument to which the father objected. From the age of 7 on, Ralf had taken piano lessons, as had his older brother and sisters. Although he cried about practicing every day, the lessons were continued. The father had brought up his children very strictly, in general, for many years. Only when they entered the stage of puberty, and he could no longer achieve anything by being strict, was he more lenient. However, then he let the reins go completely slack and resigned. He appeared that way to others, too: he walked stooped over and seemed depressive.

One must picture the mother as being completely different from the father. She was 13 years younger than her husband. In her nature, she appeared to be very emotionally oriented and differentiated in her interpersonal relations. She had been frequently compelled to intervene as an intermediary between the father and Ralf. The diagnosis indicated a relatively undisturbed relationship between her and the patient. Biographically, however, it is especially interesting that during Ralf's first years of life, she had once been at the point of wanting to leave the family because of marital stress. Now, however, the parents seemed to stand totally together again. During the initial interview, the father lovingly stroked his wife's hands. Together, they were worried about Ralf. They thankfully took the opportunity of treatment for their son and the accompanying consultations for themselves.

Ralf, too, was prepared to come to the therapy sessions. After having been previously prepared for it, Ralf arrived punctually for his first GAI session. His appearance, short and stocky, reminded one of his father's. Now, after not having attended school for six full months, his father had presented him with an ultimatum: within three weeks, he was to propose a plan with regard to how he thought he would set up his future. He had taken a job in the wrapping section of a department store but had quit after three days. His experience at this job had been the occasion for him to argue with his father again about the "exploitation of the workers by the capitalists."

In his first sessions, it became very obvious how strongly pronounced Ralf's wish was to escape from everyday reality.

First Session: December 4, 1966

The first session of GAI was intentionally limited to 10 minutes in order to first get an impression of the patient's imaginative capacity.

He was able to imagine a meadow immediately: "The sun is shining . . . and now I walk across the meadow . . . I come to a forest . . . everything is very beautiful here."

Urged to observe the edge of the woods and see what emerged from it, he recognized a light blond 18-year-old girl. She approached him directly and became tender. She was so active—he had always wished for a girl like that. Later on, he described how he walked behind a bush with her and they had intimate relations. And afterward she gave him a car as a present. He, Ralf, sat behind the steering wheel and drove to a restaurant with her, where they had a big dinner together.

Second Session: December 11, 1966

Imaginings in GAI: Ralf is playing music together with other young people somewhere in the south in a meadow by the sea. He describes a joyful vacation scene. They swim, play, and "laze around." He is living with his friends in a commune. It is nice that one does not have to pick up after oneself there. He finds a young American particularly likable. Unfortunately, his appearance reminds Ralf of a friend who lost his life through drugs.

Ralf also spends the next day of the vacation with the group by the sea. However, today dark clouds draw near. The ocean is getting stormy. The lighthouse guard, an old, seemingly good-natured man, comes. He warns the boys about the storm and invites them to come with him into the tower. While they are having coffee with him, a hailstorm is in progress outside.

After the session, Ralf seems obviously relaxed and his originally gloomy mood has disappeared. He is surprised how well he was able to picture the images.

Fifth Session: January 2, 1967

GAI: Today, Ralf is in a seemingly desolate, autumnal meadow. It is endlessly broad, totally rooted up by mice and covered with mounds made by moles. It could start to rain at any moment. Birds screech through the air. He is totally alone. Suddenly, he sees a bulldozer, 10 meters tall, coming toward him. He cannot detect a driver on top of it. Instead, he feels more and more threatened. Finally, he is able to escape into a bunker, which apparently dates from World War II. But the bulldozer has lowered its gigantic power shovel and is about to dispose of the bunker. In this emergency, he decides to raise a white flag. In the same instant, a small, bald-headed man jumps down from the vehicle. Gesticulating wildly, his face bright red and inflamed with rage, he drives Ralf out of the meadow.

In safety again, Ralf decides to climb a 20,000-meter-high mountain with his brother. In the mountains, he encounters an ancient hermit who entertains both of them hospitably. At the end of the session, when Ralf again returns to the meadow, he discovers a warning sign: "No trespassing!" A bulldozer of totally normal size is driving across the lot. Its driver explains in friendly terms that the meadow is still strewn with mines from the war. He shows Ralf and his brother a path by which they can leave safely.

SEVENTH SESSION: JANUARY 9, 1967

Today, Ralf imagines that he is a giant. He estimates his height to be 400 meters and reports, "The people are as small as ants! I have to watch out that I don't trample anyone to death with my big feet!"

However, in Norway, he finds an equally large person with whom he makes friends. This person is dressed completely in hides, and only his eyes are visible. Ralf and his friend wanted to go fishing together. They already see one fish. It is several hundred meters long, very old, and can see out of only one eye. They want to kill it immediately, but all the arrows bounce off it. Even bombs, which the fish eats packed inside of large pieces of cheese, only cause a soft gurgling in its stomach.

Later on, when Ralf runs into distress at sea while on a boat trip with his friend, the big fish reappears. It takes both of them onto its back and carries them to shore. Out of gratitude they feed it seals. Now, the fish itself suddenly comes onto land and speaks. It has a deck of skat cards wedged between its gills and wants to play with the boys. They agree immediately. To keep the animal from drying out, they cover its body with large pieces of ice. The fish wins all the games. Suddenly, they notice that it is cheating. It has five additional aces hidden under its chin. The animal turns all red in the face from embarrassment at being discovered. It begins to speak again: It is very old now and has to die soon. They will never see it again. Tears are rolling down the fish's cheeks now. Ralf and the snowperson cried too. They fondle the fish again. Then, it swims away into the North Sea.

Now both continue playing alone. The snowperson takes off his headgear. A very old man with black hair emerges from under it. He says that he will also die soon. A little later, when Ralf wakes up after sleeping for a short time, he is alone. He finds a note with a message regarding the friend: He has met his death while riding in his boat. Ralf is very sad. He wants to go to the place where the fish and the snowperson died. For this purpose, he obtains several airplanes. He ties them together, sits down on top of them, and flies north. Having reached the North Sea, he suddenly sees a gigantic hole. All of the water of the sea runs into it, as if into a pipe. A little later, however, he discovers land and is able to land in a meadow. He finds a city where the people speak German. Suddenly, he becomes uneasy about his own size. He wants to assume normal size again and goes to a physician for this reason. The doctor tells him that he can best be helped with a gigantic washing machine. However, it is very hot and the treatment can be dangerous to his life.

In the discussion that follows, Ralf is very reflective. The fish reminds him of his father ("all arrows bounce off" him, too) and of the grandfather, who often used to play cards with Ralf.

Intermediate Reflections

The somewhat confusing and quickly changing contents of the first sessions signal, first of all, superficial wishful fantasies, which are disposed of (i.e., acted out). Gloomy clouds and hailstorms announce a threatening situation. As was often the case later on, a protective father figure intervenes at this point (transference feelings for the therapist?). The actual problem is addressed in the fifth session. The desolate meadow and the threat presented by the bulldozer impressively symbolize Ralf's infantile anxieties toward his father. The driver of the bulldozer, who tramples over everything, is occupied with removing the traces of the war. He is angry that Ralf has dared to enter this (his) territory. Does his father perhaps have something to hide here? Does not this relic from the war also symbolize the scene of his earlier marital strife with his mother? The subsequent encounter with the figure of the bald-headed "little man" deactivates the situation considerably as a perhaps realistically colored reduction of the father, who had been superelevated in an infantile narcissistic manner, as the bulldozer driver. The grandfather figure is just as conciliatory as the figure of a brother, who soon appears, leads the way out of danger, and would like to leave the explosive mines lying where they are.

In the seventh session, Ralf imagines himself as a powerful giant in the sense of illusory pubertal omnipotence (narcissism). Again, he finds a brotherly figure of the same kind, with whom he can become friends. They catch an immense fish, which is invulnerable and helps them later on. Helpers from the sea of the unconscious appear without any specific explanation being given. Soon, however, the animal shows "deceitful traits" during the game. It dies, literally out of shame, mourning for itself. The helping powers of the deep unconscious are thus humanized (image of the grandfather?) and are also not unproblematic. The brotherly snowperson is also old and wants to die. Mourning and loss lead Ralf to the ends of the earth and thus, at the same time, to people and to the physician, who suggests a rigorous, hot treatment in the washing machine in order to free Ralf from the conceptions of immense size and thus lead him back to man, whom he had previously perceived as an ant. The transference feelings for the therapist and the therapy, thereby Ralf's motive for treatment, become more and more obvious. He would like to be cleared of personal flaws, too, the motif of purification as a frequent initial dream in an analytic therapy. The sucking hole as a representation of the maternal world and the unconscious, into which it threatens to pull him, also becomes evident. In fact, Ralf went through a severe crisis in the weeks that followed and repeatedly did not appear for his sessions. When he did return, progress became obvious.

Tenth Session: March 13, 1967

Today, Ralf imagines a sandy desert instead of a meadow. Several steel pegs, painted green, jut out of the earth, spaced a meter apart. He states excitedly, "Now there's a car coming . . . all the tires are flat . . . an older man gets out . . . he swears terribly . . . now he takes the wheels off . . . he wants to drive to the street this way . . . he makes his family get out of the car . . . he gives his wife a rope to put around her neck, she is supposed to pull the car . . . the three children have to push now . . . the man tyrannizes them all a lot."

Ralf is very upset. He wants to leave the area and climb up a hunter's lookout in the nearby woods. But then, he suddenly notices that there is a deer on top of the lookout. It is holding a rifle and binoculars between its legs. A hunter comes along the path with his dog. Then, the deer shoots the dog dead. The hunter runs away.

Ralf himself has not been in danger. However, he is startled by an elephant that suddenly approaches him and uproots trees. Then, the animal steps on one of the steel pegs by mistake and falls over stunned. When Ralf frees the elephant from the steel peg, it wakes up again immediately. It thanks Ralf profusely and spontaneously invites him to visit its elephant family. Ralf feels very well there. Now, he learns that the meadow had been completely green until five years ago. Then, it had been passed over to government ownership. The steel pegs were inserted to keep the animals away. Because of a suggestion by the therapist, Ralf now pours water onto the sandy desert. The elephants help him do this and dig canals using their trunks. It immediately begins to turn green everywhere. At the end of the session, all of the steel pegs have disappeared. The animals are very happy about that.

In the discussion afterward, Ralf associates the following with deer: tenderness, love, girlfriend, family, etc. He reports in passing that he has met a new girlfriend in the past week.

THIRTEENTH SESSION: APRIL 12, 1967

Ralf sees a meadow in which many animals are located. A hunter drives up in a green Volkswagen. The animals flee. The hunter becomes enraged, because they have trampled over his grass. As he does not have his rifle with him, the animals come out of their hiding places again and listen to the screaming. But the hunter raves more and more. His face turns bright red. He becomes so incensed that his own dog is startled by him and bites him. When the hunter thrashes the dog for this, the other animals come to its aid. They surround the hunter from all sides and push him to the ground. But his shouting only becomes louder. It attracts more and more animals. Even flocks of birds circle in the air and observe him. Suddenly, a helicopter appears in the sky, too. Grzimek, the animal researcher, is sitting in it. He is of the opinion that in reality, the hunter is a "screaming ape." He injects him with a tranquilizer and transports him to the nearest zoo. There, he is locked up in a cage in the monkey house. The animals follow along to the zoo to see the screamer there. His appearance has changed: Now he has pig's ears, violet hair, and a hide over his entire body. When he wakes up, he starts screaming again immediately. When no one can say anything to him because of his own screaming, they make use of a megaphone. He promptly screams back using a megaphone, too. The sound waves are so strong that the building crashes down instantaneously. The screaming ape flees to the city. The fire department and the police are out in full force. Ralf is after him, too. But then, the screamer suddenly assumes human form again. He drives away in a VW, as if nothing had happened. The next day, the hunter reads in the newspaper what a colossal sensation he created, and he is very enthused about it.

In the follow-up discussion, Ralf reports that the screaming had clearly reminded him of his father, who had become enraged in a similar fashion when he learned of Ralf's poor school grades.

Fourteenth Session: April 28, 1967

For the first time, Ralf pictures images from his real surroundings: He finds himself in the garden of his parents' home. His father is in the process of tending to the lawn. Suddenly, an airplane appears in the sky; it is spraying poison. Father and son run into the house. The grass turns brown. The father's face becomes bright red with rage. He immediately files a complaint at the office of the interior. There, he learns that a seed company is having all ugly lawns destroyed. They assure the father that the damage will be compensated for. However, he does not intend to wait that long. For that reason, he attaches a plow to his car. He drives it like a wild man over the grass and cultivates it. In his overzealousness, he destroys the adjoining flower beds. As a result, he gets into a quarrel with his wife, who takes care of the flowers. However, his zeal for work cannot be slowed down. He sows seed and waters it at the same time. The grass shoots up at the same pace at which he is working. It is already a half meter—two meters—five meters tall. The father is lost in it. He cannot free himself anymore and screams for help. Ralf cuts a path through to him with a machete. In the process, he also inadvertently cuts off some of his father's hair. Then, both of them emerge from the grass. The father goes to his desk without thanking Ralf for freeing him. He continues working as if nothing had happened.

In the follow-up discussion, Ralf reports that in reality, his father is very clumsy, too. He can neither hammer a nail straight into the wall nor operate machines. He always hides behind his desk, not only in difficult situations but also at other times.

Fifteenth Session: May 11, 1967

Although today, when asked, Ralf cannot remember any of the images from the last session, he again imagines tall grass. Again, he cuts a path through it with a knife. In the process, he strikes a man, who is about 30 years old and fishing on dry land. He is a very peculiar tramp. Strangely enough, he is sitting in a boat that has four wheels and is located in the middle of the meadow. Not a worm, but a fish is hanging from the fishing rod as bait. And he does not want to catch fish, but earthworms.

(On the supposition that this figure probably represents a split-off portion of the patient's ego, the therapist's intention was to establish contact, as closely as possible, between Ralf and this eccentric fellow.) Therefore, the tramp is invited to dinner. The two are sitting in the kitchen of a nearby farmhouse. The farmer serves the food. The tramp is very hungry. However, he eats alternately from the foods that have been placed before him and then from his worms. Then, he wants to leave again. In the course of the session, however, Ralf succeeds in repeatedly inviting the tramp to dinner (Leuner's principle of feeding). Finally, the tramp moves to the farm totally. Once he is induced to follow behind the farmer's plow, he is enthusiastic about the many worms that he finds in the process. He buys himself a plow, and the farmer gives him a piece of land, so that he can plow it for himself. Ralf himself helps with the fieldwork. At the end of the session, he plants a tree and waters it.

Intermediate Reflections

Changes in the direction of positive development become more and more frequently obvious within the sessions. The negative father image thereby also loses its threatening quality and becomes more human, and the father acts, in part, like a caricature of his real behavior.

The tenth session runs like a play in three acts. First, the patient is confronted with a negative father symbol. Many details are amazingly suggestive of the real family: Ralf also has three siblings. Five years before, there had also been a period that had, in reality, been of decisive importance. At that time, the father had been exposed to the hostilities of the press, and at the same time, several grandparents had died. In Ralf's family, the mother must also have been the person, who, as in his pictured image, "cleared up the mess" in cases of emergency. The second section of this catathymic drama is characterized by the aggressions directed against the father figure in the form of the hunter. Ralf's associations with the symbolic figure of the deer lead one to suspect that he is hoping for reinforcement against paternal authority from his new girlfriend.

In the thirteenth session a confrontation with the negative father image recurs, but now it is in a form closer to consciousness. The driver of the car and the hunter are identical from the start (the father also drives a VW). Man and animal gather together as if to a common crescendo, expelling the danger in order to hold in check the father image, which has been degraded to a screaming ape. Characteristically, this episode likewise creates a sensation in the press (see above). As in reality, however, they cannot find anything against him.

The fourteenth session takes a form even closer to consciousness, when Ralf spontaneously imagines his father in person for the first time. Only gradually does he acknowledge his aggressions toward his father. He has him experience failure in his garden work. He "inadvertently" cuts off his hair (which, in reality, the father no longer has). But Ralf receives no thanks for rescuing his father.

The fifteenth session reveals that the unconscious event has many more layers than at first seemed to be the case. At first, Ralf appears to make a correction, "Look, the one in the tall grass is not only your father. He is also a part of yourself that you have to accept." The homeless tramp, who is frustrated with his fishing, is received as a guest (assimilation of a split-off part of the personality). He, like Ralf himself, finally becomes united in rural labor through cultivating the earth—an activity recollective of the late Faust, which is productive and is experienced as being pleasant. The figure of the farmer probably reflects the giving role of the benevolent therapist. This development is symbolized in a dream that Ralf had immediately afterward. A spliced television tower becomes whole again, while Ralf is playing music in a meadow. The assimilation of previously split-off parts of his personality continues to be the central theme of the future sessions.

Eighteenth Session: June 26, 1967

Today, Ralf is in a meadow under water. At the moment, he is diving. At the bottom of the sea, he suddenly discovers an apartment in which 20 people are living. They have retreated from the earth because there are always so many wars and so much hatred on earth. Ralf is welcomed by them in a friendly manner and participates in their excursions in the ocean environment. He finds out that they have never seen the sun. He is gradually successful in motivating all 20 people to surface at least once. But they always dive into the depths again immediately. Further encouragement to surface is given. Thus, they examine a tree with fruit and a grain field. Then, they visit the weekly market in the city for the first time. They have never before seen so many things at once and are immensely astonished. They buy a donkey from a dealer and keep it on an island. On the island, they also build houses for themselves. Finally, they build a lighthouse and construct a port. To protect themselves from unwelcome visitors, they secure the shore with barbed wire. Now, they themselves perceive their old apartments on the bottom of the sea as uncomfortable. Now, they only rarely dive down there. Only when military maneuvers are held at sea do they retreat into their old apartments for 10 days. However, then they come up again and inhabit the island along with Ralf.

Twenty-first Session: July 12, 1967

A confrontation with the mother had been planned for today's session. However, after only a short time the father appears again and is sitting at the steering wheel of a VW. The family is on a Sunday excursion. They order cake in a café. As Ralf begins to feel quite comfortable, his father suddenly reproaches him for having skipped school again. At first, Ralf defends himself. Then, however, he gets upset to the point of throwing the cake in his father's face. Consequently, he gets a slap in the face from his father. He is nevertheless very satisfied that he was finally able to give vent to his aggressions so clearly. This feeling of relief continues after the session, too. It is the first time that he has been able to express so much rage toward his father.

Twenty-second Session: July 17, 1967

Ralf sees a man in diving gear in the meadow. To his amazement, he is diving in a pond that is only a few square meters in size. The man is around 60 years old and makes an unusually eccentric impression. For dinner, he picks himself a few berries. With his harpoon, he pulls only little fish out of the pond—not to eat, but to stuff.

The man is very annoyed when Ralf brings up a really big fish out of the pond without having diving gear. His eyes bulge out with rage like a frog's eyes. Angry, he lies down to sleep in the middle of the meadow. Ralf wakes him as a thunderstorm is coming up. The man gets angry again about this. He says that the rain would not have bothered him at all, because he is wearing a diving suit. Then, he disappears into the forest.

In the village, Ralf learns that the diver comes from the "loony bin," a mental institution. He was wounded by a shot in the head during the war. Since then, he has believed that the war is still going on and has retreated into a cabin in the woods. Ralf succeeds in making friends with this man. Suddenly, however, he discovers that the inhabitants of the village do not talk with him anymore and even spit on the ground in front of him. Neither by spitting back nor through a sensible discussion is he able to influence the rejecting attitude of the people. However, this changes when he brings the diver along to the village. Now, the people correct their opinion about him and the diver. The diver can move into an apartment in the village. Ralf, however, says good-bye to him and takes to the road.

In the follow-up discussion, it strikes Ralf that the diver isolated himself in a way similar to his father. A "war wound" was discussed in detail, which in its way is common to the father as a symbolic figure. Today, the difficult conflictual situations, which had been experienced five years before, were again expressed as had been done earlier. Colleagues in Ralf's class had passed around a daily newspaper when Ralf's father had been publicly attacked and had whispered that Ralf was "the son of that one there." At that point, Ralf had to decide on whose side he wanted to stand internally, his father's or that of "public opinion." Although Ralf was convinced that his father was objectively innocent, he had at that time decided against him outwardly.

INTERMEDIATE REFLECTIONS

Ralf seems to anticipate that he can restructure his attitude toward his father in the GAI sessions by assimilating split-off parts of his personality. In the image of the inhabitants of the sea floor, they have, so to speak, retreated from active life. Making contact with the world, which the therapist intended, and their settlement on the island, still isolated from man, are stations of the "return." However, prominent signs of masculinity are already set with the lighthouse and the harbor. Ralf's actual contact with his environment also improved during this period. He reported that he had talked about the stressful events from five years ago with his parents beforehand, a subject that had been taboo for many years. Because of school vacation, an extended break in the treatment took place. The therapeutic process, which was developing intensively, was scarcely impaired by this.

TWENTY-THIRD SESSION: OCTOBER 25, 1967

Today, Ralf imagines a house in the meadow. He is married, has two children, and lives there. At the moment, he is in the cellar playing music with his friend. Just now, his wife calls him to eat. They are having chicken. Everything tastes good. Suddenly, however, the friend begins to eat more and more. He devours a loaf of bread in two bites. This friend is very obtrusive in general. He calls up at all times of the day and night, which particularly disturbs Ralf's wife. Even now, after the meal, the friend does not want to leave. Ralf hopes to get rid of him with the comment that he does not have any more cake to go along with the coffee. However, that comment cannot embarrass the friend. He has brought along cookies in his car, two hospital packages, each holding 1,000 units. He eats 10 of them at a time. In so doing, he gets so many crumbs on the floor that Ralf has to ask him to sweep them up. Afterward, the friend does drive home, but he phones again immediately and asks if they want to have dinner together that evening. Ralf, however, prefers to go somewhere alone with his wife and puts the friend off until the next day. Later, he reports that he goes to a physician with his friend. The doctor prescribes a "rehabilitative treatment of withdrawal" in Bad Mergentheim. After that, the friend lives very sensibly.

In the follow-up discussion, Ralf reports that he is very much looking forward to having a family of his own someday.

TWENTY-FIFTH SESSION: NOVEMBER 16, 1967

Ralf imagines a motorcycle race. He crosses a swampy area along with 20 other drivers. Beyond that is a village, which is the finish of the race. Whereas Ralf is making good progress, more and more of his rivals are sinking into the mud. Cursing and crying, they leave their vehicles stuck in the swamp and walk to the village. Ralf makes good headway following the bumpy track of a bulldozer. Almost all of the competitors have already withdrawn. Ralf has only three opponents left. He is in third place. When he leaves the bulldozer's track, however, he is even able to take the lead. The victory flags have already been raised at the entrance to the village, flags of Germany, of the village, and of Ralf's family. The church bells ring at the victory celebration. Prizes are to be distributed. However, he and the other two decline. Therefore, the fourth one, who arrives five hours late, gets the prize. The minister is present, too. He is particularly looking for contact with the people. For this reason, he often goes to the pub and has a drink. Above all, however, he is an enthusiastic motorcycle driver. He practices regularly with his machine in the swampy area. Since he wears a black cowl, one does not even notice if he gets thoroughly splashed. He has oddly constructed the entrance to his church so that all the people can drive in on their motorcycles and do not need to get off them during the church service. The minister can even drive his motorcycle up to the pulpit. The organist plays the organ without leaving his vehicle. The contents of the sermons are exclusively about the motorcycle. For example, the minister recites a passage from the Bible such as "Let there be light" and explains, in what follows, the electrical system of the motorcycle, including headlights, brakelights, and ignition. Since he has been doing that, the church is full again.

However, not only the church but the entire life of the village is determined by motorcycles. Other vehicles are not permitted to be driven in the village and are turned back at the local boundary. Public transportation consists exclusively of motorcycle tandems. Up to 10 persons can sit on them at one time.

Ralf reports that a local motorcycle racetrack is being constructed there. The swamp is being drained, and, as in former times, a recreational lake is being made.

In the follow-up discussion, Ralf thinks that for him the motorcycle could have something to do with freedom and sexuality. With regard to the bulldozer track, Ralf recalls his father's expectation that his children should stay in his "track." Ralf himself has noticed that the number of winners corresponds to the number of siblings in his family.

Twenty-sixth Session: November 20, 1967

Ralf imagines that he is on an excursion train taking part in a mystery trip. The engineer is a 70-year-old man. They are playing music on the train. Ralf is playing in the band, too. The ride goes straight into a woods. There, everybody gets out and goes for a walk. Suddenly, there is a big shock; the train drives on without taking even one of those on the excursion. Nobody knows where Ralf is. Only with effort does Ralf succeed in finding a way out of the woods. He procures a bus and brings the lost excursionists out of the woods.

Some time ago, Ralf had reported having taken an entrance examination to a conservatory but that he had failed. He had no more great hopes in this direction. Therefore, he applied for a position as drummer in approximately 20 bands both at home and abroad.

Intermediate Reflections

Ralf is dealing with his future more and more. He is thinking about settling down at some point in the future. His own oral, pretentious attitude (impressively symbolized by the gluttonous friend) seems to him to be obstructive in this. In fantasy, he deals with the ambitious goal of coming in first. He can thereby fulfill his father's expectations. Thus, he seems to be conscious of the fact that he has already "missed the train" on his former mystery trip—his unplanned concept of life—during which he was playing music. This is apparently a small insight into reality. The engineer, 70 years old, probably represents the grandfather, who died five years before, and from whom Ralf had received the encouragement to play the drums. However, Ralf rescues the helpless excursionists and is master of the total situation through improvisation.

The twenty-fifth session still deserves special consideration. One interpretation, although somewhat radical, could be the following. The four winners of the race are the four children in the patient's family. The swampy area corresponds symbolically to Ralf's tendency toward regression into areas of anal drives, into which, however, he does not lapse in this imagining after all. The motorcycle minister embodies the ideal or the desired father (on the level of the subject, a strange amalgamation of a superego authority, into which suppressed expansive impulses enter in excess, perhaps in connection with the figure of the therapist). The rivals, who are eliminated in the race, correspond to former competitive classmates; Ralf would like to "show them up for once." In the village, a fantasied society becomes recognizable, the subculture of the "motorcycle worshipers" and fanatics (ideal formation in adolescence).

The motorcycle had often occupied a prominent position in Ralf's dreams and fantasies. Here, it not only embodies the wish for freedom and the possibility for aggressive abreaction but also seems to stand strongly for pubertally experienced sexuality: The driver can squeeze the cycle between his legs and lie down on top. It radiates heat from the "hot pipes." He can show off "in traffic" (German *Verkehr* = "traffic" and "sexual intercourse"). Its owner can forget himself in a dangerously thrilling trip at full speed, as in an intimate high also experienced "dangerously." One must also recall the girl who sits on the pillion seat and clings to the driver without having to be expressly encouraged to do so. The figure of speech to "get in gear" with a girl fits here. Crossing the swampy area with the motorcycle could symbolize that Ralf is seeking to master the danger of the libidinal bind to his mother. Detaching himself from his parental home and settling down manifest themselves more and more strongly in his wishful fantasies.

Twenty-ninth Session: December 11, 1967

At first, Ralf reports that he does not want to repeat the entrance examination for the conservatory. His mother is very sad about that. I advise him to confront her in GAI. Ralf is able to imagine his mother immediately. She is working in the kitchen. She has tears in her eyes. Ralf is of the opinion that she is crying only because he has applied to foreign bands and wants to leave home. Once before, five years ago, she had been similarly shocked. At that time, she had just visited her mother, who was in the hospital. With tears in her eyes, she had related how grandma had said good-bye to her from her deathbed. Then, however, Ralf lets his imagination go: Suddenly, his father appears and takes Ralf to task because he still has not begun any education. Ralf explodes into a rage against his parents and his brother. First, he throws a plate of soup in his mother's face. Without tending to her, he goes into his brother's room. There, he sets fire to his brother's drums and demolishes his car. However, Ralf vents his strongest emotions in his father's room. He sets the curtains on fire, tips over the desk and the bookcase, sprays the books with a gardenhose, slashes the table lamp with a knife, and tears down the ceiling light. He chops apart the desk with a hatchet and smashes the windowpanes. Ralf determines with satisfaction that all of them—father, mother, and brother—are hiding helplessly in a corner and crying. Then, he steals the savings bankbooks and moves to Munich.

In the follow-up discussion, Ralf shows a strong feeling of relief. He laments that his father has spent all these years behind his desk. He has never really had any time for Ralf.

Thirtieth Session: December 18, 1967

Ralf again imagines his parents. His father is now very old and has sunken eyes, and his hands have become sinewy. He himself talks about his aging. Later, Ralf meets his parents in a café. They want to sit down with him. However, he declines. Instead, he starts a conversation with a young woman, makes friends with her, accompanies her home, and then sleeps with her. This woman already has a child.

Thirty-First Session: January 8, 1968

Ralf's hopes of getting a position as a result of his applications vacillate from hour to hour. He is willing to imagine a view of his future: He is accepted into a band in Cologne. They are playing in a commune at the outskirts of the city. When Ralf enters the room, an audience is also present, who are talking about society. He ascertains critically that the drummer is beating his instrument very aggressively, as if he were intending to demolish a car. Strangely enough, he is exactly the same age as Ralf. He, Ralf, does not have anything to "report" in this group yet, because he has just recently arrived. He does not want to play music with the others either. Instead, he is flirting with a girl, with whom he soon becomes intimate friends. He marries her, and they have two children. He is looking for a position with a better band.

In the follow-up discussion, Ralf himself is surprised that not the music, but a girlfriend and family, had been the focus. His need for attachment and protection within the framework of a new (his own) family becomes obvious. At the same time, a backlog need is probably also being expressed, after failures and disappointments in his own family have burdened him. Here, he is at the same time fantasizing himself as the founder of a new family.

Thirty-Fourth Session: February 1, 1968

Ralf reports that for the first time he has made friends with a younger girl. Contact with the older girlfriend, whom he had actually wanted to marry at one point, has been broken off completely.

In GAI, Ralf imagines a bunker, from which music is resounding. The bunker has neither windows nor doors and is surrounded by barbed wire. Ralf can get into the bunker only by using great effort. There, he meets five musicians, who are all approximately 20 years old. They are playing in a terribly boring fashion. The room is extremely unfriendly, damp and cold, and it smells musty. Spiders are crawling everywhere. The light is very dim. The musicians are so sluggish that one after another finally falls asleep. Ralf uses all means to try to wake them up—in vain. Among other attempts, Ralf pours a bucket of water into the drummer's face. The drummer thinks that it is hard liquor and only mutters sleepily, "Another gin, please!" Then, Ralf decides to knock an opening through to the bunker, so that sunlight can get in. The hole has to be punched from the inside outward. If one were to punch the hole from the outside inward, the walls would collapse onto the sleeping musicians. He goes to work with a hammer. As the first sunbeams penetrate into the bunker, the musicians suddenly wake up. They are not accustomed to sunlight anymore. Therefore, they have to wear protective glasses. They themselves help with the subsequent tearing down of the bunker. When Ralf asks them, they cannot say where they came from. They have neither a clock nor a calendar. They have lived on only bananas and oranges.

In the course of the session, Ralf succeeds in inducing the musicians to walk to the nearby village. There, he finds an apartment for them. Now, they play music in the townhall regularly. They gradually learn to like normal food a lot. When the bunker again becomes the topic of conversation, they only laugh.

Thirty-fifth Session: March 28, 1968

Ralf has got a position as a musician. Today is his last session. In GAI, he imagines once again in retrospect the path that he has traveled during treatment. In a jungle, he lived with a lion family. He says good-bye to the animals, crosses a large river, and enters into a civilized area. There, he befriends a farmer, finds a job working for him, and moves into his own room.

INTERMEDIATE REFLECTIONS

In this phase of treatment, Ralf first allows himself to express massive aggressions toward his parents, which can probably be interpreted as revengeful impulses released against everything that they did not do for him. He also gives his brother his share, and not without feeling triumphant. Finally, he renders his father powerless and demonstrates to him and to himself his father's impotence in the figure of an old man who is becoming helpless.

The theme "contact with a female partner" assumes an increasingly definite form. Instead of devoting himself to music, he finally settles down. More realistically, he now commits himself to a girlfriend of his own age, instead of his former maternal girlfriend (an example of practical behavioral correction after previous dry rehearsals in GAI).

The bunker theme with the confined, tired musicians is parallel to the inhabitants of the sea floor during the earlier period: split-off parts of the person living apart from the world, exhausted, and having become one-sided. They are thankful for being led back into life, to say the least (a typically pubertal problem area, which frequently emerges in a way similar to this within this age group).

The lion family can perhaps be interpreted symbolically as follows: independent, animal life, yet without commitments, in which all impulses, particularly the aggressive ones, can be given free play. This world is abandoned in favor of a realistic, human life, which is filled with obligations.

RESULTS OF THERAPY

Retesting was performed at the end of the treatment (requiring 35 sessions) of this severely disturbed adolescent, who came from a difficult family background. The Rorschach test showed a clear decrease in the basic depressive mood. The portion of aggressive answers had diminished. The capacity to solve and master conflicts was considerably more pronounced than before treatment. The progress in the Freiburger Personality Inventory (FPI) (62) was even more clear. The standard value of depressiveness decreased from 7 to 5; the values for composure and strivings for dominance improved respectively from 4 to 2 points.

The complaints described by Ralf had developed in the following way: anxieties no longer appeared, suicidal wishes had completely disappeared. He no longer noted teeth-gnashing at night; headaches and diarrhea did not appear any more either. His relations to his father were significantly improved. Ralf had even written a poem for his father on his birthday.

During the past half year, Ralf had attended trade school regularly. To be able to fulfill the requirements for a high school diploma, he participated in a course of studies within the scope of adult education classes over a period of a few months. His parents noticed that he had become more active. For example, he had renovated the apartment on his own, while they were away. He took dancing lessons. In addition, he passed the test for his driver's license. At the time of his eighteenth birthday, he had an appointment with a new band at a French seaside resort. He attended to all the formalities of moving to that place on his own. His parents loaned him money to purchase his own car for transporting his instruments. Ralf had paid off his debts approximately six months later. He reported that he was financing a building loan contract with his savings. We traced Ralf's development over the four years after treatment was terminated. Up to that time, he had been steadily employed with bands and had remained free of any major psychopathological problems or psychosomatic symptoms.

Commentary (Leuner)

In commenting on this very clear contribution, which is easily comprehensible in the presentation of the psychodynamics, I would like to restrict myself to a few summarizing remarks. Some lines of development can, in turn, be contrasted with one another on the basis of the contents and the therapeutic process of this severely disturbed patient. The marked confrontation with the paternal-masculine world stands in the forefront. On the one hand, a gradual process of detachment from the father occurred and, on the other hand, an increasing development of his own identity in the masculine realm. The detachment process was conveyed by working through the narcissistic superelevation of the father image and of the patient's own self. Finding his identity was promoted by the mobilizing of helping figures ("pacemakers"), who were apparently modeled, on the one hand, on the corrective emotional experiences in the transference to the therapist and, on the other hand, on a reference to the image of the grandfather. The mobilization of repressed and split-off parts of the person was particularly impressive within the framework of finding his own identity. A strengthening of the mature ego was connected with this. At another point, we described an analogous process, which is apparently characteristic in the psychotherapy of adolescents. That case concerned a giant, representing the narcissistic grandiose self, that had withdrawn into a cave and built his own realm there.

Another developmental line can be shown in the repeated dry rehearsals in the various stages of development, through which the following were expressed: the setting free of aggressive impulses; the release from revengeful feelings, in part, in the form of identifying with the aggressor (father); and finally, the development of more strongly expansive-motoric and sexual impulses, in the sense of satisfying archaic needs, as was previously mentioned. The conveyors of these impulses were, in part, fantasies of a subculture, in which they were both encouraged and at the same time embedded. Through this case study, it becomes clear what a psychodynamically significant part the formation of a subculture has for this age group.

An additional line was the unfolding of the gradual development of partner attachments, not to be exclusively understood as sexual, with the fulfillment of wishes for tenderness, love, and lasting fellowship in the form of settling down. A prospective tendency was being developed at the same time. It was surely conveyed in part by the family of origin; however, the wish for a new beginning (Balint, 2) in a domestic fellowship, which corresponded to his expectations of human relationships, was realized much more strongly here in fantasy.

One criticism needing comment is the reader's desire to know more about the manifestation of the transference in the individual phases of treatment, even though this can be surmised from the contents at some points. However, the reader would also like to know the details of the therapist's countertransference feelings, which were apparently of a strongly positive character and were conveyed by a brotherly, caring stance.

The question raised in the psychotherapy of adults as to the character-altering effect of a therapy, a question particularly justified in the case of a treatment requiring only 35 sessions, seems to be mistakenly posed in the present case. The adolescent age group finds itself by its very nature on a tempestuous developmental course with regard to character, in which spontaneous changes are the rule. Therefore, the question should read, instead, whether the therapy fundamentally contributed to emphatically promoting the existing latent developmental tendencies and, within the framework of this developmental course, decisively contributing to resolving neurotic and character deformities. In this way, the patient could lose his neurotic traits in the "melting-pot of puberty" and arrive at his individual development with far fewer limitations and with the goal of finding the best possible identity. In studying this clinical history, one can scarcely doubt that this goal was attained. That Ralf still has further important steps to take in his development, now on his own resources, does not reduce the significance of this therapy for him but instead corresponds to performing the further developmental canon as is expected. He owes the fact that he is now capable of developing farther, relatively free of disorders, to this therapy, in which he was a delightfully cooperative patient.

10

Guided Affective Imagery in the Treatment of an 8-Year-Old Neurotic Boy

GÜNTHER HORN

When Klaus had his first appointment at the age of 7 years, 11 months, he seemed tall and strong for his age. He went into the observation room with me without being afraid. However, his shyness and his very serious-looking face were striking in his behavior. His dark hair and dark eyes as well as his soft voice intensified this impression even more.

A series of numerous symptoms had given rise to the first appointment: extremely aggressive outbursts at home, excessive motoric impulses, a tendency toward isolation, eating disturbances, and anxieties relating to school, with a sharp drop in achievement.

Psychodiagnostically, a considerable retardation of psychic development was apparent, particularly with oral and anal fixation. Extreme anxieties and obsessive-compulsive traits were in evidence. His relationship to his mother seemed to be seriously disturbed. Moreover, his extremely minimal readiness for action was conspicuous in the projective tests.

In Klaus's development, a series of momentous factors fostering disturbances was noteworthy. His mother's early attitude toward her child had been rejecting. His mother had never become more closely acquainted with his father, who was a foreigner who had been only a passing acquaintance during Carnival. At the time of Klaus's birth, his mother was living in the country with her parents and grandparents. She was plagued by the constant fear that her pregnancy might be discovered. However, she had succeeded in keeping it a secret until the day of his birth. She had presumably laced up her body for this purpose.

Klaus's mother had suffered from depression not only during the pregnancy but even in her childhood. She herself recognized that she had been overstrained "in every respect" with the birth of the boy. Therefore, Klaus was taken care of by his grandparents and by a nursemaid in his first years. His mother continued working in her parents' grocery store. Whereas she gave Klaus scarcely any attention at that time, he was allegedly "spoiled excessively" by the grandparents. Then, however, grave traumas occurred: At the age of 2 ½ years he was the first of all the family members to find his great-grandmother dead. She had suffered a stroke while in the bathroom. Not long afterward, when he was 3 years old, his mother married. He was transplanted into the newly formed marriage and into another place of residence. From then on, he suffered very severely from homesickness for his grandparents. The situation grew even worse when a half-sister was born, and he forfeited the rest of the central position in the family to which he had become accustomed.

His stepfather could not help Klaus in this situation. After having discontinued his study of art and many years' service in the army, he got married and began a new course of study at a teachers' college. He considered himself "very inhibited." He felt totally helpless in the face of his stepson's aggressions. Disappointed with military service, he had consistently forbidden Klaus to play with rubber dartguns and capguns. He was very dependent on his parents and could not defend himself against their reproach of having married "beneath his class" and "a woman with a child." Neither could he have any influence on the distinct favoritism shown to the little daughter as opposed to Klaus. Family tensions, which had been going on for years between his wife and his parents, burdened the new marriage severely. Klaus obviously suffered in this tension-filled domestic atmosphere, and consequently he lacked an upbringing that offered him security.

The therapy began by admitting Klaus and his mother into a mother–child group. However, Klaus could not make contact with the other children and their parents. I thereupon decided to take Klaus into individual treatment with guided affective imagery (GAI). In contrast to his behavior during the projective test procedures, Klaus showed an unusual capacity and readiness for imagination in GAI. I conducted a total of 24 sessions of GAI, each lasting 20 minutes. Even in the first GAI session, it became apparent how Klaus perceived his environment: sad, depressing, and hopeless; in a word, he showed all the signs of a depressive state. It was raining in the imaginary meadow. He did not have an umbrella with him. With the help of the therapist, he finally succeeded in finding shelter under a protecting roof. Then, he complained that he was still getting wet. Cars drove through puddles and spattered him. I had the impression that the "bad weather" was related to his perception of his parental home, because in order to make it more friendly, he took a bouquet of flowers home in GAI. Then the image changed and suddenly the sun began to shine. A brisk little brook was there. Squirrels, a hunter, and a dog came out of the woods.

The abundance of the images expressed in further GAI sessions was so great that it is difficult to select the most prominent ones for this presentation. One remarkable aspect of the treatment was that he took up relatively "acute" problems at the beginning and, in the course of the sessions that followed, those that dated further and further back.

To begin with, it became clear how he was currently perceiving school: as confining and boring. He had very strong motoric needs. Thus, he imagined dreaming that he was racing across the meadows on a motorcycle. As he awakened within the dream, he was already thinking about how bored he would be in school again. It was cold on the way to school. Snow was on the ground.

Most of the other contents of the first sessions also referred to current problems. From the third session on, however, more and more irrational images emerged—evidence that for the first time the regression was now going back to the magical stage of life. It was that stage of life during which Klaus experienced the severe traumas mentioned above. In his imagination, he first went through confrontations with his stepsister, which grew more and more intense. His extremely ambivalent relationship with her became perceptible. He imagined, for example, how he was playing with her peacefully and, in the next moment, had her incur injury: They each took turns going down a long slide into a big lake. When his sister landed in the water, a group of hungry whales was awaiting her.

The main confrontation with her took place in symbols, which gave rise to the suspicion that they had played sexual games together.

Thus, Klaus often imagined that he slept in the same room with her. Then, suddenly, his sister called out that her bed was on fire. Then, the fire department regularly squirted water into it. In this connection, the ever-recurring image of a hostilely perceived stork was revealing. It picked up his sister and carried her into its nest.

Until the eighth session, it was mainly the stepfather who had appeared in Klaus's fantasy. He had known him since he was 4 years old. In the ninth session, the regression extended into the more distant past. Now, his mother became the focus. Here, a scene arose that can be interpreted as a birth fantasy. First, he imagined cows who told jokes to each other, tickled each other, and laughed. A little later Klaus was swimming in a big pond with his clothes on. He let himself glide over a big waterfall and into the depths. His clothes had shrunk because of the water. Now, he had "doll's clothes" on his body. Then, he suddenly encountered a woman, whom he gradually recognized as his mother. On account of her poor eyesight, she unfortunately could not detect that he was wearing doll's clothes. So he went with her to an optometrist. However, he discovered that she had something wrong with her appendix. Another doctor operated on her.

A few months before his stepsister's birth, Klaus had experienced his great-grandmother's death. It was impressive that in the tenth session, only two sessions after the birth image, he now experienced the deaths of each of his grandparents and parents in succession in his daydreaming. The grandmother died of "heart pain," the grandfather of ailments of the head and legs. His father and mother lost their "lives" through shock about that. In this scene, a part of the early childhood separation from his parents seemed to take place simultaneously, as it otherwise usually occurs during the age of defiance. In his imagination, the house in which his parents had lived was being remodeled. A new ego boundary was expressed graphically in a 20-meter-high wall, which surrounded the entire house. In addition, several soldiers with guns in a panzer drove up for his protection. A "terrible slaughter" developed during a battle with other soldiers. Now, all of those emotions seemed to manifest themselves that Klaus had not dared to express until then out of fear of loss of love.

In the following sessions, the regression continued consistently. The anal quality of his aggressions was unmistakable in the following: Klaus mentioned "shooting," "shitting," and swimming in a lake and the ocean (intrauterine fantasies) in close connection. In the eleventh session, for example, ducks encouraged him to swim in a lake with them. When he declined because he did not have a bathing suit along, the ducks put a bomb in his rectum. Klaus did not offer any resistance. Instead he was delighted when it exploded and the water in the lake splashed. Then, he imagined a cannon, which he used to shoot at the ducks. Suddenly, he reported that one of the ducks was transformed into a "woman with blond hair, 3 years old." It was Anke, his girlfriend, whom he would have liked to have as a sister.

Somewhat later, in the twelfth session, the theme of separation became evident in a new variation for the first time. From that point on, he imagined almost regularly that he was riding on the ocean, and with the help of the police, his mother was trying in vain to bring him home. His ambivalent relationship to his parental home was impressively exhibited in this fantasy, because at night he secretly crept into his room in order to sleep at home.

Characteristically, the oral fixation and the process of working it through appeared in images only after sexual and anal themes had already been worked through. Klaus suffered from an eating disorder. Foods often nauseated him. At other people's homes, he did not eat anything. In the fifteenth session, he imagined grass that had grown particularly tall. The stork was sitting on top of it. It was so big that it could watch God eating. There were two empty chairs next to God: one for Jesus, the other for the Holy Ghost. Unfortunately, God did not enjoy being seen by the stork. In the next session, Klaus reported that he often did not feel hungry at mealtimes. Then, he always saw a snake on his plate, which made him nauseous. He did not dare to tell his father about it. In the subsequent imaginings, the oral theme stood clearly in the forefront. Klaus imagined, for example, that he was planting all kinds of vegetables next to his house and setting up an area for grilling. Gigantic fish, crocodiles, dragons, and other animals with large mouths appeared in his fantasies. Scenes from the land of milk and honey were imagined, in which passive-accepting behavior such as eating, watching television, and sleeping played a major part. Sections from the word-for-word transcript of the eighteenth session are meant to illustrate these drive tendencies that now arose:

(Therapist: "Where does the water come from?")—Klaus; "From the ocean."—("Does it taste good?")—"Like sauerkraut . . . and then I go home again—and then I tell my grandma, 'Grandma, I was gone for two days.'"—("Oh yes! What does grandma look like? Is she there now?")—"Hm, now—now says—now I noticed, that she is not there—that there's just a figure standing there . . . she's not there anymore . . . "—("What did she look like?")—"The way she used to look, with white hair and wearing glasses."—("And did she like you? Or did she want to be alone?")—"She li—lik—liked me."—("Are you sad now, that she is not there anymore?")—"Nn . . . then I sleep in the captain's room . . . and sleep more and the next day I don't wake up, as long as I sleep . . . I'm dreaming something . . . that I'm having breakfast and the bird comes, it drinks all the coffee . . . I said: 'But you have to drink water, but not me, we aren't birds.' Then it said, 'I'm a bird and you are people, so then I'll drink coffee here' . . . then I say: 'Now listen you, OK! you smart aleck! Birds drink—ah—water and people drink coffee and not the other way around.' Then it says, 'Keep your mouth shut! I'll drink coffee now, and if it's not all right with you, you can leave' . . . "—("Does the bird remind you of anything?")—"Mhm. Of my other bird, the one I bought."—("Mhm. When did you get it?")—"When my grandma was still alive . . . and now I want to have a real breakfast . . . and then I'll go to the movies . . . and then I'll go home again and then I'm so very tired again and sleep . . . can't stop sleeping, then I offer the captain a seat, turn—on the television. And I cook something special on the grill, a chicken . . . "—("Did you ever see a person like this captain before?")—"Nn—only my grandpa, only my grandpa—and then—I get it ready and then we eat it and then we go inside and sleep some more. It's evening."—("Does the captain sleep in your room with you?")—"He sleeps in the same bed with me."—("Did you ever sleep with someone like that?")—"Mhm, with my sis—with my sister."—("Was that nice, too? Or is this nicer now?")—"It's nicer now—because there used to be two beds next to each other, and now there is only one bed there. And then we sleep and we sleep. This time I don't dream anything. Only the crocodile screams and screams and opens its mouth ."

In the following sessions, a new theme took an increasingly broader scope. A spirit of adventure more and more strongly replaced the blatant sleeping and eating needs. Age-appropriate wishes for expansion—as, for example, taking possession of islands along with battles as a knight and a soldier—were prevalent. Mother symbols, on the other hand, receded into the background. Father symbols, which were experienced positively, appeared. In the twenty-first session, Klaus imagined how an emperor traveled across the ocean on a ship. Klaus sat at the helm and navigated. That he identified himself very closely with the emperor is shown by the fact that his family name corresponded almost exactly with the designation *emperor*.

In his last GAI session Klaus also imagined a trip across the ocean. Everything that, in his opinion, he needed to survive (as, for example, stove, table, chairs, tools), he had with him. The mood in which he delivered his report seemed decidedly happy. In comparison to the first session at the beginning of treatment, the absence of depressive contents was unmistakable.

Also, his other behavior during the therapy sessions had changed. In conversation, he now seemed to be substantially freer. The pictures that he regularly drew after the GAI sessions were diffusely drawn and had little structure until the eighteenth session, from which excerpts were quoted. Now, they assumed a substantially clearer expression. His last picture was a self-portrait. Klaus painted himself as a soldier in a yellow and blue uniform, having a husky stature and armed with two swords.

Summary

The development of the treatment can be presented as follows: The focus was the 24 individual sessions. They took place from March to November 1973. Their contents consisted of approximately 20-minute GAI sessions with subsequent drawing of the previously imagined pictures. Klaus's imaginings were not interpreted but were discussed in a participative manner. Moreover, I occasionally talked with him about his concrete difficulties. With a view to the apparently still necessary and gradually proceeding separation from me, I took Klaus into a small group that I was leading, after the individual sessions were terminated. He participated in the group once a week until March 1974. The focus here was more on pedagogically oriented games and dialogues. Klaus's behavior in the group was substantially more outgoing than in the mother–child group mentioned previously. At the termination of this treatment, I could not observe any difficulties with separation.

Parallel to Klaus's individual and group sessions, I held counseling sessions with his parents at approximately six-week intervals. The primary goal of these sessions was to change their attitude toward Klaus. His father's participation in a self-awareness group for parents from January to March 1974 had the same purpose. We did not run a psychological control test at the end of Klaus's treatment. However, one can conclude that a substantial improvement of the psychological disorder had taken place based on the complete disappearance of the symptoms by the end of treatment. After Klaus' thirteenth session, his parents reported an improvement in his isolated behavior for the first time. Now Klaus frequently roamed about with other boys. At this point in treatment, Klaus imagined for the first time that he was riding on the ocean, and that his mother was trying in vain to bring him home again with the help of the police. In a discussion with the parents after the sixteenth session, further improvements became known: His eating disorder had disappeared. Oral contents had been the focus of the preceding GAI sessions.

He had dropped his original tendency to eat with only a knife and a fork. Now he could also eat at other people's houses without feeling nauseous. His sensitivity to dirt had disappeared. He could now play in the mud like other children.

The symptoms relating to school were the most persistent. His fear of school did not disappear until very near the end of the individual treatment. His achievement in school was still very unsteady until that point and had, as a whole, improved only slightly. His achievement improved only during the following months, during which time his father changed his attitude—within the scope of the self-awareness group—toward achievement, toward a choice of occupation, and toward Klaus. At the time, he was studying at a teachers' college and suffering from intense feelings of inferiority.

Consultations were held with the parents until November 1974. No relapses could be observed up to that time. Klaus's achievement in school was also satisfactory at that time.

COMMENTARY (LEUNER)

Based on the symptoms, a severely neurotic clinical picture was depicted in the case of Klaus: aggressive outbursts, intensified motoric drive, a tendency toward isolation, eating disorders, school phobia, sharp decline in achievement, serious retardation in psychological development, extreme anxieties, and obsessive-compulsive traits. The anamnestic data correspond with this picture: being an unwanted child, rejection by the depressive mother, traumatism through the discovery of the grandmother's death, transplantation into the mother's new marriage, the birth of a half-sister who was strongly preferred later on, continuing tensions in the new family, and the rejection of the child by the "stepgrandparents." Accordingly, the description is one of a severely depressed, motorically inhibited child who avoided contact with others and whose really helpless parents could contribute little to the mastery of the situation.

Therapeutic attempts in a mother–child group failed. Despite the prevalence of depressive contents in the meadow, the first GAI session already signaled possibilities for the better with a positive phenomenon of change after a bouquet of flowers was brought into the house. The development of the treatment, which required only 24 sessions despite the child's severe disturbance, was characterized by three developmental criteria, in which a certain internal logical consistency can be recognized: (1) positive retrogression in the working through of acute problems and those dating further back; (2) a progressive working-through of the typical developmental phases (even if in an unsystematic sequence) and expansive liberation by means of experiencing oral, anal, and sexual fantasies of liberation; and (3) gradual separation from important referential figures through the release of aggressive impulses without triggering off feelings of guilt. The release of such archaic impulses and their gratification occurred regularly within deep regression and symbolic representation, whereby a special role was given to the expansively aggressive representations and the changeover from the maternal to the paternal symbolism, as well as the identification with the paternal images.

The symptoms improved with the same logical consistency as that with which the symptoms and the anamnesis stood in relationship to the therapeutic process. The depressive contents and behaviors diminished; after 13 sessions, the contact problems disappeared, after 16, the eating disorders, and finally, as the last and most burdensome symptom, the school-related fears disappeared and his school achievement improved.

It is therapeutically fitting and corresponds to general experience in child psychotherapy that besides the individual therapy of such a severely disturbed child, who was in the stressful milieu of obviously neurotically inhibited, immature parents, a therapy for the parents was offered for prophylactic reasons to prevent a relapse and with the aim of improving the child's further developmental possibilities.

However, these accompanying measures were limited and, according to general experience in child psychotherapy, would not alone (i.e., without the individual therapy in GAI as the focus) have sufficed to stimulate a child so severely disturbed with regard to his behavior and milieu, in the manner described. Several questions remain unaswered, above all, the boy's transference to the therapist, the parents' attitude toward the therapy, the cause of their attitudinal change toward the child, and lastly, also the therapist's countertransference to the child, who, in view of all the attention given to the therapeutic process, presented him with a "nice case." Considering all of these things, the case of this 8-year-old severely disturbed boy illustrates the intensive therapeutic effect that the procedure has on children and the possibility through GAI of directly encouraging emotional developmental processes and the mastery of ambivalent attachments to the persons in charge of his upbringing and thereby attaining behavioral modifications "without coercion." As far as the clinical success of the treatment can be judged, this occurred in the present case within a short-term therapy of 24 sessions despite the severely damaged milieu of the child.

11

Guided Affective Imagery in the Short-Term Therapy of an Eyelid Tic

INGE KLEMPERER

Tics resist psychotherapeutic efforts with particular obstinacy. The eyelid tic of a 10-year-old girl, on which neither medication nor autogenic training had any influence, ceased after the fourth treatment session with guided affective imagery (GAI) with no symptom substitution. Tics occasionally disappear spontaneously, too. In this case, the course of therapy suggests a connection between the giving up of the symptom and the treatment.

Lieselotte was the oldest of three children. From what her mother stated, her development proceeded "super smoothly." Even as a baby, she was very calm. She was toilet-trained early, had no phase of defiance, never caused any difficulties whatsoever, and, after starting school very early, brought home only very good grades. When Lieselotte was 6 years old, her father was forced to retire early. After an operation on his thyroid gland, he suffered from paralysis, spastic attacks, and high blood pressure. His continual presence inhibited the motoric activity of the children. They always had to be considerate and quiet. At this time, an eyelid tic first appeared in Lieselotte for one day.

Lieselotte's mother, who had attended high school, felt that she had gone down in the social scale through her marriage to a laborer. Now, after her husband's retirement, she directed all her interest and her energy toward her daughters' upbringing, particularly the oldest daughter, so that her daughters should attain what had been denied her.

At the age of 10, Lieselotte had spent the summer vacation at her grandmother's house. Her mother reported indignantly that her sister-in-law, who lived with the grandmother, allowed the child to stay up later than was usual at home. She returned home "totally nervous," and since that time, her eyes had been "twitching." Three years after treatment had been terminated, I learned that the mother had been pregnant during that period.

As already indicated, I treated Lieselotte's tic with medication and autogenic training for four months. The symptoms improved temporarily to some extent, only to subsequently emerge more severely.

During that time, I had become acquainted with GAI and, based on the short duration of the symptoms and the dynamics inherent in the child's dreams at night, I decided on an attempt at treatment using this procedure.

Eight sessions were held, with treatment being conducted once a week (and a three-week vacation break after the sixth session). After the eighth session, Lieselotte failed to return, without giving any notice.

I conducted the treatment in accordance with Leuner's standard motifs. In the following, I quote only the most significant passages of the treatment.

First Session (Meadow Motif)

The dramatic beginning of the first hour illustrated Lieselotte's basic mood:

Two rabbits are playing in a brown, trampled meadow surrounded by a fence. A fox creeps up, grabs one of them, bites it at the nape of the neck, and sets it aside to catch the other one. That one has, however, already run into the woods. The fox grabs the first one again, drags it into its den, and eats it up.

The main motifs of the later catathymic panorama already emerge within an initial dream during the rest of the session: a border, behind it a foreign country, desert, extreme heat, thirst; a path that is interrupted by obstacles; and also wandering, resting, and making oneself comfortable. Lieselotte needs my help in difficult situations, for example, in searching for water in the desert.

At the end of this first session, the meadow changed (phenomenon of transformation in accordance with Leuner,) now it was green and had colorful flowers.

Second Session (Pursuing the Course of a Stream, Mountain Motif)

This time, a narrow stream emerges from the large, green meadow. At first, Lieselotte follows it, but then it oozes away in the forest and does not emerge again, although she searches for it for a long time. Then she swims in a lake, climbs a steep, rocky hill, and looks down on a highway with many cars.

During the second GAI, Lieselotte was very restless. She was silent a lot but showed an active play of features and lively arm and leg movements, which induced me not to intervene with suggestions but to communicate to her the experience of patient and composed waiting.

Third and Fourth Sessions

The third and fourth sessions were very turbulent and apparently particularly important for the course of treatment. In these sessions, the scenes of action were constantly changing. Lieselotte moved around in the image as if she had been set loose, "and on and on . . . " was the tenor of these sessions. The lively and colorful images developed in the midst of deep relaxation. The strong emotional participation of the young patient now became increasingly obvious. To give an impression of this pressing fullness, I quote the third session word for word:

Meadow, many houses with red roofs and large windows. I run up and down the path and back to the meadow, into a house, up and down the steps and back to the meadow again. Black, white, and brown dogs are jumping around there, digging and playing with each other, and then they run away. I go into the attic of a house, onto the roof, have a look at the landscape: city, trees, gardens, and forest. I come down from the roof, go for a walk on the sidewalk, on and on into the city, out into the woods. Many animals: deer, foxes, even a tame stag. On along the path into the woods, out of the woods, into a town. Farms, simple houses, streets, paths, restaurants, horses, cows, chickens, dogs, cats, pigs. Out of the town onto a path in the fields. A farmer is mowing grain in the fields. Further onto a street. Lots of cars. Next to the street, there is a railroad track. A train approaches and disappears. Further along the field path into a woods, into the meadow. The dogs are still there, biting and scratching themselves now. Now the weather is bad, it's raining. I go into the entryway of a house to protect myself from the rain. The meadow is squishy, I slip. Further along the path, in another direction. A factory, black, smoky, it stinks terribly, that's the way to school. I come to a river, absolutely clear, deep water, the shore is grassy and dry with colorful flowers. The river gets narrower, stops in the meadow. It has nice, soft grass with white, red, and yellow flowers. The weather is sunny and dry. I pick a really big bouquet. I walk on, come to a mountain but would not like to climb it, walk to the foot of the mountain, up to the border, through a gate in the fence into a very strange country, with yellow people with braids, slit eyes, and strange handwriting. I see a village with small houses; I lie down, relax, the sun shines, fall asleep, feel at ease, but a little bit cool from the ground. Wake up again, walk further to a cliff with a cave, into it, and walk to the end. Skeletons of people are hanging there. I get scared, out of the cave again fast. Further to a forest with very small trees; I can look over them. Through to a forest with big trees, whose crowns I cannot see, to a village with small houses and tiny little people, who run away from me. I walk into the forest so that I will not make them afraid. I walk to the border. A bear approaches me, is peaceful, wants to play with me; we romp about and knock each other down. His fur is soft and brown. He has a big head, long tail, very small paws, and hare's ears. His head looks red. But he grunts just like other bears do, goes along with me. We encounter many big and little cats, lions, tigers, and other wild cats. I run away together with the bear; the big cats go on. I come to the sea, lie down on the warm, white sand, stay lying there for quite a while. A rowboat, a green boat, with lilac-colored oars and yellow on the inside. Ask people on the beach if the boat belongs to them. They say no. Climb in and go to sea. See many boats. I do not like it, row back. A man comes along whom I did not ask before, because he looked so fierce. He scolds me, and I run away fast. The bear is still with me, runs behind me. The beach gets smaller and smaller, because the tide is coming in. I go up a staircase and along the top, further and further. A city with nothing but skyscrapers, no normal cars on the streets, just fast racing cars. I do not like it, it is too loud. Over the border into a forest. A hill, not very high, the path is dry, not very steep, not rocky, just at the edge. From above, I look down on the landscape, city with skyscrapers. Down to a brook with trout (I suggest following it first to the source and later downstream): a little opening out of which clear water bubbles up. Downstream it oozes away in a field that is black from having burned. (I encourage her to see whether the brook appears again.) She walks on and on, it does not appear for the present. The ground is reddish brown with some tufts of grass. She continues to search in all directions. Finally, the brook does appear again to the right, gets wider and wider; then it is a very wide river with muddy shores without any grass, dirty water, flows quickly into a city, through it along a street. Horse-drawn carriages with figures on top of them, many disguised people: princesses, dogs, cats, Chinese people, cowboys, Indians, ghosts—further on, watch a Carnival parade, it is over; the city is at an end: (Encouraged by me) Back to a meadow, tired from wandering, I lie down, comfortably tired, warm, sunny, cannot fall asleep, just lying there like that. Very warm.

The richness of imagination and the pace of the productions were also striking in the further development but were never again as excessive as in this third session.

In the fourth-session GAI:

The flowery meadow is surrounded by a street, on which many cars are driving but cannot drive down from the street. Behind this there is a big wall, and without encouragement Lieselotte looks for a ladder to get over it and does find one. The river is wide and has clear, calmly flowing water. Now excursions into fantastic, exotic worlds ensue. She leaps over borders, undergoes diverse adventures: desert, heat, cannibals, bandits who shoot at her with bows and arrows. Finally, she looks for a hotel and gets Room 371 on the fourth floor. But the fourth floor only goes to Room 370, the fifth begins with 372. She goes to the doorman, who shows her Room 371 on the third floor. She has an argument with the doorman about this, then falls asleep and wakes up again 20 years later. She looks in the mirror, has long hair and long fingernails. A man who is 65 years old is standing in the room. He is a sheriff, who had previously appeared in her imaginings, and with whom she talks about her experiences.

Only two images were selected from the variety of details, which indicated a change of behavior in GAI: the escape from the anxiety-producing boat owner in the third session and entering into the argument with the doorman during the search for the room in the fourth session.

Subsequent Sessions

In the ensuing sessions, the radius of action grew continually larger in the imaginings. Some of the situations, near at hand and far away, which were passed through in a seemingly chaotic succession in the preceding sessions, were now experienced by Lieselotte in breadth. The number of difficult situations increased to which she faced up courageously and which she solved independently. Examples:

> In the fifth GAI, an excursion with her grandmother and her aunt, on which occasion she takes along a basket full of things to eat. When she loses her way on the long hike, she searches for the way home without my assistance. An excursion in the sixth GAI alone with an unknown goal, blanket, sandwich, and thermos on her back. Ghosts frighten her. At my suggestion, she appeases them with candy, which she puts into their big, wide mouths. In the next hour, she finds the key to the trunk, inside which the ghosts are locked. On a journey at sea, she is driven off course onto an island. Without becoming panicky, she builds a hut, goes fishing and hunting, and is of the opinion that she will make it OK, as Robinson Crusoe was able to do it, too.

From these few examples, the growing feeling of her own strength and independence in the course of treatment became evident.

Another problem was first intimated in the third GAI in the image of the cave. Full of fear, Lieselotte at that time fled out of the cave, which she had dared to enter in order to explore it. In the fifth GAI, she found herself in a world with very small people who had big heads or fat stomachs. Out of curiosity, she tried to explore this world. In a later session, Lieselotte swam down a wide, peacefully flowing river on a raft. A whale swallowed her, along with the raft. It was dark in his belly. She took the raft apart and made a fire with it in the whale's belly to find out what it looked like inside. The smoke made the whale cough; and he spit her out again.

In the eighth session of GAI, exotic and familiar motifs were again intermingled—desert and meadow:

> Lieselotte wanders, then climbs a steep, rocky mountain with difficulty and perseverance, relaxes with satisfaction, crosses the ocean on a raft, which she had made herself, lands at a city and is glad that there is "solid ground in front of her door again." Then the treatment is broken off abruptly. Lieselotte no longer came to therapy; the eyelid tic had already disappeared after the fourth session.

I saw Lieselotte again three years later. She had remained free of symptoms. The difficulties for which her mother sent her to me now argued in favor of a normal development, the adolescent's age-appropriate attempts at separation.

Thoughts about the Course of Treatment

The immense richness of imagination, particularly in the third and fourth sessions, certainly expressed a release of the child's past extensive motoric obstruction. She had apparently been extensively and aggressively inhibited from a very early age, along with a hysterical component in her character structure. Of course, the flightiness also signaled defense, but I had the impression that above all else, it brought effective relief from the intense intrapsychic pressure and, through dry rehearsals in the imagination, allowed motorically expansive impulses to be released and experienced. Related to this was the conquest of "new worlds" (i.e., previously not assimilated, because they were split-off, affectively emotional areas of her own person). The gratification value of these often-hastily-produced fantasies signaled at the same time the oncoming release of tension through them. Starting points could also be repeatedly found for treating the confining problems relating to authority and assertion connected with this (for example, the confrontation with the doorman), as well as initial steps toward finding her identity, such as wandering alone and looking for the room.

As mentioned at the outset, the appearance of the symptoms coincided temporally with the start of her mother's pregnancy. Fantasies of pregnancy and birth had more significance for Lieselotte than I could recognize at the beginning.

After the abrupt discontinuation, I at first had the feeling that the treatment had not been terminated. Some time later, I looked through the transcripts again and realized that the therapy had found at least a preliminary termination. An indication: the adventure in the vein of Robinson Crusoe in the seventh session ended with Lieselotte's being awakened by her mother. The mother said that it was all only a dream, and that that was a good thing, because in reality she certainly would not have endured all that. Immediately afterward, Lieselotte imagined, "I am on vacation, I am going swimming, I swim myself free."

COMMENTARY (LEUNER)

The treatment of tics and analogous conversion symptoms by means of GAI is ordinarily not easy and runs into limits, as in other forms of therapy as well. For this reason, the present case must be seen as having met with particularly fortunate circumstances insofar as the young patient was able to release her motorically expansive impulses in GAI relatively early in the first session and to experience them in fantasy. In the second session, this is expressed more through motoric restlessness and, logically, in the third, through a richness of imagined activities all the way up to aggressive impulses, as in the image of the biting dogs and cats and the mud and dirt fantasies, as well as in the factory representation. Characteristic of a spontaneous development in GAI are the "pacemakers" (i.e., guiding helpers), who strengthen the therapeutic principle of the ego, as in this case the symbolic figure of the bear. The expansive needs continue to be experienced fully in the ocean voyage, which, however, is not carried to the fullest; in the city with the fast racing cars (here, there are also limits because of the noise); and finally, in the fourth session, in a similar manner with the scaling of the wall, which leads the patient into foreign (i.e., kept away from the ego) regions of exotic worlds and adventures. The radius of action is strikingly expanded even within the sequence. Therefore, the phenomenon of satisfying the archaic need for motoric and thus also affective-emotional expansion and the assimilation of suppressed aspects of the ego pervades the treatment as the main thread. The confrontation with superego figures and their conquest, is already indicated in the text and is just as logically consistent as the attempt to find the new identity based on the new experiences in GAI. The fact that the symptom disappeared after the session with the most intensive expansion, after the fourth GAI, also lies within the framework of this logical consistency.

One cannot assume that an extensive restructuring of character took place within this child during the treatment, which lasted eight sessions. However, for this age group, corrective emotional experiences with the release of regressive (i.e., archaic) impulses, which were somaticized in the form of the eyelid tic, represent a great deal of relief and—as we see—liberation from the symptom. The prospective tendencies also become obvious in the next to the last session in the apt comment, "I swim myself free." Some separation from both internal and external confining authorities is doubtlessly connected with this image. The course of therapy shows that solutions can be attained quickly in a therapeutic manner through the "opening of the gates to the unconscious" using GAI along with the positive transference to a friendly, giving therapist. Only a longer follow-up observation period can show to what extent the solutions are lasting, which we have occasionally seen to be definitely the case. The influence of the child-rearing milieu will persist unless the relevant referential figures are, in turn, included in the therapeutic process by way of counseling, group work, or other similar kinds of therapy, as has already been repeatedly pointed out.

References

1. Aichhorn, A. (1973): Verwahrloste Jugend. Bern-Stuttgart (Huber) [7]1971.
2. Balint, M. (1970): Therapeutische Aspekte der Regression. Die Theorie der Grundstörung. Stuttgart (Klett).
3. Battegay, R. (1970): Angst und Sein. Stuttgart (Hippokrates).
4. Beck, D., Lambelet, L. (1972): Resultate der psychoanalytisch orientierten Kurztherapie bei 30 psychosomatisch Kranken. Psyche 26, 265.
5. Beck, M. (1968): Rehabilitation eines chronischen Trinkers mit der Methode des Katathymen Bilderlebens. Praxis Psychotherapie 12, 97.
6. Bellak, L. (Ed.). (1973): Conceptual and methodological problems in psychoanalysis. Annals of the New York Academy of Science, 73, 4.
7. Breuer, K., Kretzer, J. (1974): Beziehungen zwischen Gesprächspsychotherapie und Katathymem Bilderleben. In: Ausgewählte Vorträge der Zentralen Weiterbildungs-Seminare für Katathymes Bilderleben (KB). Göttingen (AGKB, Selbstverlag).
8. Bruch, H. (1965): The psychiatric diagnosis of Anorexia nervosa. In Meyer and Feldmann.
9. Bruch, H. (1976): Psychotherapie bei Magersucht und Fettsucht des Kindes. In G. Biermann, (Ed.).: Handbuch der Kinderpsychotherapie, Ergänzungsband. München-Basel (Reinhardt).
10. Desoille, R. (1945): Le rêve éveillé, Psyche 1, 37 (Paris).
11. Erikson, E. H. (1968): Identity—Youth and crisis. New York (Norton).
12. Fleck, L. (1969): Pubertätsmagersucht des jungen Mädchen und ihre Behandlung. In G. Biermann, (Ed.).: Handbuch der Kinderpsychotherapie, Bd. II. München-Basel (Reinhardt) [4]1976.
13. Frahm, H. (1965): Ergebnisse einer systematisch durchgeführten somatisch orientierten Behandlungsform bei Kranken mit Anorexia nervosa. In: Meyer and Feldmann.
14. Freud, A. (1946): The ego and the mechanisms of defense. New York (International Universities Press).
15. Freud, S. (1953): The interpretation of dreams (Complete works). In J. Strachey (Ed.). New York (Hogarth Press).
16. Fromm-Reichmann, F. (1956): Principles of intensive psychotherapy. Chicago (University of Chicago Press).
17. Gallwitz, A. (1965): Versuch einer experimentellen Erfassung des body Image bei weiblichen Magersüchtigen. In: Meyer and Feldmann.

18. Grünholz, G. (1971): Vom LSD zur Selbsthypnose in "psychedelischer" Erfahrung. *Z. Psychoth. u. Med. Psychol. 21.*
19. Happich, C. (1932): Das Bildbewußtsein als Ansatzstelle psychischer Behandlung. *Zbl. Psychotherapie 5,* 633.
20. Heiss, R. (1956): Allgemeine Tiefenpsychologie. Bern-Stuttgart (Huber).
21. Holfeld, H., Leuner, H. (1969): "Vatermord" als zentraler Konflikt einer psychogenen Psychose. *Nervenarzt 40,* 203.
22. Klessmann, E. (1973): Therapiemöglichkeiten bei jüngeren Drogenkonsumenten in der Erziehungsberatungsstelle einer Kleinstadt. *Praxis Kinderpsychol. 27,* 513.
23. Koch, W. (1969): Kurztherapie einer zwangsstrukturierten Neurose mit dem Katathymen Bilderleben. *Z. Psychother. u. Med. Psychol. 19,* 187.
24. Kohut, H. (1968): The psychoanalytic therapy of narcissistic personality disorders. Paper presented at the meeting of the Third Freud Anniversary Lecture of the Psychoanalytic Association, New York, May 20. (Published in German).
25. Kornadt, H.-J. (1960): Der Zusammenhang zwischen allgemeinem Anspruchsniveau und bestimmten Merkmalen bildhafter Vorstellungen. Bonn (Internat. Psychol. Kongreß).
26. Kretschmer, E. (1922): Medizinische Psychologie. Stuttgart (Thieme) [14]1975.
27. Leuner, H. (1953): Über die jugendpsychiatrische Bedeutung von Reifestörungen. *Z. f. Kinderpsychiatrie 20,* 12.
28. Leuner, H. (1954): Kontrolle der Symbolinterpretation im experimentellen Verfahren. *Z. Psychother. u. Med. Psychol. 4,* 201.
29. Leuner, H. (1955): Experimentelles Katathymes Bilderleben als ein klinisches Verfahren der Psychotherapie. *Z. Psychother. u. Med. Psychol. 5,* 185 u. 233.
30. Leuner, H. (1955): Symbolkonfrontation, ein nicht-interpretierendes Vorgehen in der Psychotherapie. *Schweiz. Arch. Neurol. u. Psychiatr. 76,* 23.
31. Leuner, H. (1957): Symboldrama, ein aktives, nicht-analysierendes Vorgehen in der Psychotherapie. *Z. Psychoth. med. Psychol. 6,* 221.
32. Leuner, H. (1959): Das Landschaftsbild als Metapher dynamischer Strukturen. In H. Stolze, (Ed.).: Der Arzt im Raum des Erlebens. München (Lehmann).
33. Leuner, H. (1960): Die Leistungen, Indikationen und Grenzen des Symboldramas. *Z. Psychother. u. Med. Psychol. 10,* 45.
34. Leuner, H. (1962): Die experimentelle Psychose. Berlin-Heidelberg (Springer).
35. Leuner, H. (1964): Das assoziative Vorgehen im Symboldrama. *Z. f. Psychother. u. Med. Psychol. 14,* 196.
36. Leuner, H. (1965): Über einige Grundprinzipien der Kurztherapie. *Z. Psychosomat. Med. u. Psychoanal. 15,* 199.
37. Leuner, H. (1967): "Kurzpsychotherapie", ihre Problematik und ihre Notwendigkeiten. *Z. Psychoth. u. Med. Psychol. 17,* 125.
38. Leuner, H. (1969): Über den Stand der Entwicklung des Katathymen Bilderlebens. *Z. f. Psychother. u. Med. Psychol. 19,* 177.
    Guided affective imagery (GAI): A method of intensive psychotherapy. *American Journal of Psychotherapy,* 1969, 23,4.
39. Leuner, H. (1969): Das Katathyme Bilderleben in der Psychotherapie von Kindern und Jugendlichen. In G. Biermann, (Ed.).: Handbuch der Kinderpsychotherapie, Bd. I. München-Basel (Reinhardt) [4]1976.
40. Leuner, H. (1983): Katathymes Bilderleben. Stuttgart (Thieme).
41. Malan, D. H. (1963): A study of brief psychotherapy. London (Tavistock).
42. Meyer, J. E., Feldmann, H. (Ed.) 1965: Anorexia nervosa. Stuttgart (Thieme).
43. Miller, A. (1971): Zur Behandlung der sogenannten narzisstischen Neurosen. *Psyche 25,* 641.

44. Richter, H. E. (1965): Die dialogische Funktion der Magersucht. In: Meyer and Feldmann.
45. Rogers, C. (1980): The way of being. Boston (Houghton Mifflin).
46. Sachsse, U. (1975): Über die Psychodynamik in der Gruppentherapie mit dem KB. In Ausgewählte Vorträge der AGKB Göttingen (AGKB, Selbstverlag).
47. Schultz, J. H. (1932): Das Autogene Training. Stuttgart (Thieme) ^{14}1973.
48. Selvini, M. (1965): Interpretation of Mental Anorexia. In: Meyer and Feldmann.
49. Silberer, H. (1912): Symbolik des Erwachens und Schwellensymbolik überhaupt. *Jb. psychoanal. Forsch. 3*, 723.
50. Specht, F., Leuner, H. (1965): Erfahrungen mit dem Symboldrama bei Kindern und Jugendlichen. *Monatsschr. Kinderheilk. 113*, 237.
51. Sperling, E. (1965): Die Magersuchtfamilie und ihre Behandlung. In: Meyer and Feldmann.
52. Stierlin, H. (1971): Das Tun des Einen ist das Tun des Anderen. Frankfurt (Suhrkamp).
52a. Stierlin, H. (1968): Short-term versus long-term psychotherapy in the light of a general theory of human relationships. *Br. J. med. Psychol. 41*, 357–367.
53. Tausch, R. (1974): Gesprächspsychotherapie. Göttingen (Hogrefe).
54. Theilgaard, A. (1965): Psychological Testing of patients with Anorexia nervosa. In: Meyer and Feldmann.
55. Thomae, H. (1961): Anorexia nervosa. Stuttgart. Bern (Klett-Huber).
56. Thomae, H. (1972): Zur Psychoanalyse der Anorexia nervosa. In: Meyer and Feldmann.
57. Tolstrup, K. (1965): Die Charakteristika der jüngeren Fälle von Anorexia nervosa. In: Meyer and Feldmann.
58. Wächter, H.-M. (1976): KB und Psychodrama. Versuch der Synthese zweier bekannter Verfahren. Vortrag auf den 6. Zentralen Weiterbildungsseminar der AGKB, (unveröff. Manuskript).
59. Wolpe, J. (1969): The practice of behavior therapy. New York (Pergamon).

Test Procedures

60. Boehm, F. (1942): Erhebung und Bearbeitung von Katamnesen. *Zbl. Psychoth. 14*, 17.
61. Brengelmann, J. C., Brengelmann, L. (1960): Deutsche Validierung von Fragebögen der Extraversion, neurotischen Tendenz und Rigidität. *Z. exp. angew. Psychol. 7*, 291.
62. Fahrenberg, H. J., Selg, H. (1970): Das Freiburger Persönlichkeitsinventär. Göttingen (Hogreve).
63. Keller, U. (1962): Neigunge-Struktur-Test (NST). Bern, Stuttgart (Huber).
64. Pudel, V. (1974): Objective measurements of psychotherapeutic results. Paper presented at the meeting of the Central Training Seminaries of the "Arbeitsgemeinschaft für Katathymes Bilderleben" (AGKB), Willingen, FRG. (Unpublished German manuscript.)
65. Raven, J. C. (1960): Guide to the standard progressive matrices, sets A,B,C,D. London.
66. Taylor, J. (1953): A personality scale of manifest anxiety. *J. Abn. soc. Psychol. 48*, 285.
67. Zenz, H. (1971): Empirische Befunde über die Giessener Fassung einer Beschwerdenliste. *Z. Psychoth. med. Psychol. 21*, 7.

Author Index

Aichhorn, A., 82, 183

Balint, M., 24, 157, 183
Battegay, R., 81, 183
Beck, M., 33, 121, 183
Boehm, 131
Brengelmann, J. C., 121
Brengelmann, L., 121
Bruch, H., 93, 183

Desoille, R., 25, 183

Fleck, L., 79, 183
Frahm, H., 92, 183
Freud, S., 16, 21, 92, 183
Fromm-Reichmann, F., 16, 183

Grünholz, G., 109, 184

Happich, C., 12, 16, 184
Heiss, R., 16, 184
Horn, G., 59, 133, 159

Keller, 112
Klemperer, I., 171
Klessmann, E., 32, 41, 77, 81, 83, 105, 122, 184
Klessmann, H.-A., 77, 80
Kohut, H., 82, 84, 184

Kornadt, H.-J., 19, 184
Kretschmer, E., 12, 17, 184

Leuner, H., 1, 9, 43, 45, 50, 58, 65, 72, 87, 93, 97, 99, 101, 102, 109, 115, 119, 144, 156, 168, 175, 180, 184

Malan, D., 93, 121, 185
Miller, A., 91, 92, 185

Pudel, V., 121

Raven, J. C., 121
Richter, H. E., 83, 185
Rogers, C., 6, 185

Schultz, J. H., 16, 185
Selvini, M., 86, 185
Silberer, H., 16, 185
Sommer, I., 5, 95
Stierlin, H., 93, 185

Tausch, R., 6, 185
Taylor, J. A., 121
Thomae, H., 92, 185
Tolstrup, K., 92, 185

Wächter, H. M., 43, 119, 185
Wolpe, J., 5, 102, 185

Subject Index

Abandonment, 18
Abreaction, 54
Access to experiential areas, 12
Accommodating attitude, 58
Accompanying feelings, 38
Active technique, 82
Actual problems, 16
Aesthetic element, 58
Age regression, 54
Aggressions, 70, 155
Aggressive abreaction, 151
   impulses, 180
   outbursts, 161
Altered state of consciousness, 4, 122
Alter-ego fantasy, 90
Amenorrhea, 81
Anal fixation, 161
   drives, 55
   sphere, 22
Animals, 18
   with large mouths, 165
Anorectic patients, 117
Anorexia nervosa, 44, 46, 49, 55, 79, 92
Anxieties, 72, 161
   after dark, 135
   at night, 67
Anxiety neurosis, 25, 31, 135
Anxious-depressive condition, 50
Apple tree, 49
Approach core problems, 93
Aquaphobia, 46, 47
"Archetypal" symbolic quality, 87
Archetypical contents, 44

Ascent of a mountain, 19
Assimilation
   of hostile contents, 36
   of split-off parts, 146
Associative procedure, 17, 30, 37, 65, 131
Asyntactical associations, 17
Autoaggressions, 71
Autogenic training, 47, 174
   at home, 110
Autosymbolism, 16

Banish with his look, 35
Basic idea, 5
Basic imaginal motifs, 13
Basis for transference, 44
Behavioral abnormalities, 46
   correction, 155
Behavior of the therapist, 38
Behavior therapy, 4, 5
Biosorbin, 80
Birth fantasy, 164
Boat, 19
   motif, 25
Bodily process in GAI, 93
Body self, 84
Bond to the parents, 22
Borderline cases, 56
Breaking through the defensive front, 82
Bridge phobia, 31
Brook, 17, 18, 47, 50, 54
Bulimia, 87

189

Castration, 57
  theme, 56
Catathymic landscape, 50
Cave, 19, 178
  motif, 25
Change of behavior, 177
Change of scene, 44
Character
  defenses, 4
  resistances, 14
Child guidance centers, 3
Childhood phobias, 95
Children's GAI, 49
Child's spontaneity, 69
Child's symbolism, 44
Chronic obstipation, 55
Clarification, 39
Clarifying dialogue, 14
Communication therapy, 117
Comparison to the psychoanalytic technique, 14
Competitive confrontation, 27
  situations, 28
Compulsive symptoms, 44
Compulsive tendencies, 67
Concept of symbolism, 103
Condensation, 17
Conflicts, 17
Confrontation, 32, 33, 98, 101, 163
  in symbols, 163
  with symbols, 30, 34
Confronting style of GAI, 90
Conquer anxiety, 34
Consciousness expansion, 117
Contact problems, 27, 131
Content represents resistances, 29
Contraindications, 46, 71
Controlled regression, 4
Conversion symptoms, 180
Core conflict, 5
Corrective emotional experiences, 180
Counseling for the parents, 112
Countertransference, 14, 30, 158, 170
Cow, 22
Creative productions, 110
Creative response, 58
Crisis intervention, 54, 58

Daydreaming, 16
Defense, 179
  against reality, 24
  mechanisms, 44
  system, 91
Defensive processes, 14, 29, 30
Defiant behavior, 66
Deficient feelings of security, 70
Depressing imaginal contents, 71
Depressive mood, 72, 123, 131, 155
  sign, 50
  state, 163
  structure, 135
Depth-psychological, 4
  psychology, 3, 6
Desoille's procedure, 25
Developmental tendency, 72
Diagnosis in child guidance, 59
Diagnostic, 28
  examples, 18
  GAI, 61, 64, 65, 67
Dialogical clinging, 83
Directive motifs, 109
Direct the symbolic processes, 14
Disempowering of the parental figures, 34
Displacement, 17
Disturbed adolescent, 133
Dog phobia, 51
Dosage of a confrontation, 29
Double-track of the therapeutic process, 34
Draw affective images, 45
Dream house, 21
Drop in achievement, 161
Drug abuser, 119
  experience in GAI, 122
Drug scene, 115
Dyadic-relational system, 92
Dynamic aspect of the neurosis, 30
Dynamic processes, 4
Dyslexia, 49

Early childhood pathological habits, 46
Eating disorder, 165
Eating disturbances, 66, 167
Edge of the woods, 17, 23, 64, 65, 137

Subject Index

Effectiveness of confrontation, 95
Ego
  boundary, 164
  control, 34
  strength, 93
Elements of the confrontation, 30
Elephant, 22, 141, 142
Emotional insights, 28
  involvement, 11
  tone, 38
Empathic attitude, 58
Encopresis, 71
Encounter with relatives, 19, 22
Enuresis, 65, 77
Expansive impulses, 180
Expressive capacity, 72
Eyelid tic, 171, 173

Face hostile figures, 35
Facial tic, 32
Failure in school, 135
Fairy-tale level, 24
Family constellation, 23
  drama, 92
  life, 21
  realities block GAI, 48
  sphere, 21
Fantasy game, 73
Father symbols, 166
Feeblemindedness, 46
Feeding
  and enriching, 33
  and reconciliation, 72, 99, 101
  and satiating, 30, 37
Feelings of omnipotence, 92
Fifteen-session therapy, 121
Figures, 28
Finger-sucking, 66
Flexibility on the therapist's part, 45
Flexible application, 44
Flowers, 17
Flower test, 64
Flowing pictorial images, 109
Flow of images, 65
Focus on a particular part, 50
Follow-up, 33
Forest, 50, 137

Free associations, 56
Free associative procedures, 29
Functional unity, 103

GAI
  as control function, 49
  in groups, 44, 105, 109
  as optimal therapeutic possibility, 83
  as short-term therapy, 40
  a surrogate for drugs, 122
Game of fantasy, 63
Gear oneself to the patient, 64
General internal tension, 68
Generous motherliness, 17
Good prognosis, 71
Group dynamic,
  109
Group fantasy, 109
Guiding the patient, 43

Headaches, 27, 66
Hebephrenic schizophrenia, 27
Height of a mountain, 19
Heterosuggestion, 109
Homosexual, 123
Hormone therapy, 81
House, 54, 70, 125, 148, 164
Hypnagogic vision, 13

Idealizing transferences, 82
Identification with the father, 20
Identity confusion, 82
Illusory expectations, 20
Image associations, 37
Imaginal consciousness,
  16, 81
Imaginal death, 71
Imaginary confrontation, 97, 102
Imagined contents, 103
Indications, contraindications, 45
Individual technical instruments, 30
Infantile anxieties, 57, 140
Infusions, 80
  and phantasies of rape, 80
Initiative in affective imagery, 70
Intellectualized children, 12
Interactional confrontation, 3

Internal confrontations, 28, 30
Internal process of change, 28
Internal psychic situation, 22
Internal "support", 110
Internal symbolic drama, 93
Interpreting the contents, 38
Interpretive references, 24
Interpret themselves, 28
Intrapsychic pressure, 179
Intrauterine fantasies, 165
Intravenous infusions, 80
Introducing the GAI, 73
Introjected images, 4
Invent a conversation, 32

Jactitation, 46

Katathymes Bilderleben, 12

Latent conflicts, 4
Level of aspiration, 19
Libidinal overtones, 26
Longing for security, 125
Long-term treatment, 39
LSD trip in GAI, 125

Magical fluids, 45
Magical stage, 163
Making the symptom comprehensible, 45
Mastery of anxiety, 93
Mature forms of ego confrontation, 28
Mature parts of the ego, 39
Meadow, 19, 47, 49, 50, 54, 64, 70, 72, 100, 125, 137, 141, 142, 169, 174, 178
  motif, 15, 55, 97
"Meaningless" repetition compulsion, 45
Memories from childhood, 22
Mental imagery, 13
Mirror transferences, 82
MMPI, 121, 129
Mobile projection, 13
Mobilization of creative attempts, 4
Mother-child symbiosis, 83
Mother symbols, 166
Mothers of anorectic patients, 83
Motif
  of the meadow, 17, 18
  of the tree, 64

Motifs of GAI, 110
Motivation for using drugs, 125
Motoric impulses, 161
Motor unrest, 68, 69
Mountain, 50, 175
Musical GAI, 115
  group GAI, 117

Nailbiting, 65, 66
Narcissistically disturbed person, 84
Narcissistic neediness, 91
Narcissistic personalities, 82
Narcissistic transference needs, 92
Need satisfaction, 122
Needs for fusion, 92
Nervous restlessness, 97
Neurotic character resistances, 39
  maladjustments, 29
New beginning, 157
  experiences, 180
  identity, 180

Objectifying, 16
Object level, 20
Observation of the edge of the woods, 19
Obsessive-compulsive traits, 161
Obstructive motifs, 21, 29, 30, 49, 89
Oedipal situation, 18
Omnipotence, 27
Operation on the symbol, 38
Oral fixation, 161, 165
  sphere, 22
  theme, 55
Organic brain syndrome, 46
Outpatient therapy, 79

Pacemaker, 48, 180
Pacify a hostile symbolic figure, 35
Paint, 110
Panoramic view, 20
Parental images, 32, 51
Passive tendencies, 124
Patients' mothers, 83
Personality change, 131
Person-centered therapy, 6, 43
Phobias, 31, 45
Phobic state, 54
Pitch of the voice, 64
Play therapy, 4, 11

Subject Index

Points of orientation, 29
Positively accented changes, 73
Positive transference, 82, 181
  transformation, 72
Preparing the child for GAI, 63
Principle of feeding, 51, 144
Private realm, 16
Procedural technique, 62
Procedure of symbolic confrontation, 34
Process of gradual clarification, 28
Progressive confrontation, 34
Projective character, 39
Projective procedure, 13
Prospective possibilities, 82
Prospective tendencies, 180
Provocation of defenses, 22
Psychodrama, 4, 43
  group, 117
Psychodynamically oriented psychotherapy, 73, 115
Psychodynamic concept, 58
  connections, 56
  dialogue, 82
Psychodynamics, 43
Psychoses, 46
Psychosomatic ailments, 21
Psychotherapeutic side effect of infusions, 80
Puberty crisis, 27
Pursuit of the course of a brook, 19
Pyrophobia, 97

Radius of action expanded, 180
Reality control, 24
Reconciliation, 30, 34, 36, 51, 72, 83, 116
  and tender embracing, 35, 37
Regression, 163, 164, 165
Regressive symptoms, 84
Rehearsals in the imagination, 179
Relaxation, 5, 13, 69, 175
  suggestions, 15
Relaxed after one session, 73
Release
  of anxieties, 36
  infantile projections, 37
  of motoric obstruction, 179
  of regressive impulses, 180
  of tension, 179

Repeated rehearsals, 157
Representatives of parental figures, 33
Repressed material, 24
Resistances, 30, 37
Restless children, 68
Revengeful impulses, 155
Revival of a night dream, 35
Richness of imagination, 177, 179
  of imagined activities, 180
Rivalry with the father, 20
Role playing, 117
Rorschach test, 155

Safety and concealment, 26
Satisfying archaic needs, 4, 157
  for motoric expansion, 180
Satisfying suppressed-drive needs, 44
Scenic images, 37
Scenotest, 49
School phobia, 66
Second "ego", 90
Self-confrontation, 3
Self-interpreting procedure, 39
Self-observation in mirrors, 84
Separation, 165
Setting, 62
Sexual anxieties, 84
Sexual symbolism, 114
Sexuality, 22
Short-term psychotherapy, 119
Short-term therapy, 12, 47, 79, 83, 97, 121, 171
Signal of therapeutic progress, 72
Significant pauses, 38
Signs of disturbance, 19
Silent group, 109
Sitting position, 45, 63
"Solo runs" at home, 55
Solution to background conflicts, 6
Special features, 41
Split-off emotional areas, 179
Split-off parts of personality, 148
  of the person, 155
Spontaneous development, 29
Stance of the therapist, 29
State
  of altered consciousness, 5
  of anxiety, 54
  of relaxation, 16
Stomach trouble, 70

Stream, 175
Strengthening of the ego, 51
Strengthening the disturbed ego, 103
Structure of anorectic patients, 49
Stutterer, 20, 25
Stuttering, 65
Substitute for the drug, 115
Substitute father figure, 36
Suggestibility, 44
Suicidal tendencies, 135
Suicide attempt, 123
Symbiotic level, 93
   unity, 85
Symboldrama, 3, 12
Symbol of the unconscious, 24
Symbolically refreshing, 21
Symbolic act of birth, 89
Symbolic character, 38
Symbolic confrontation, 37, 53, 93, 97
Symbolic dream, 5
Symbolic figures, 32
Symbolic language, 43, 93
Symbolic manifestations, 5, 19
Symbolic mastery, 5
Symbolic representations, 4
Symbols of depth psychology, 29

Tape transcription, 63
Technical conditions, 62
   implementation of GAI, 15
Technique
   of feeding and satiating, 35
   of symbolic drama, 34
Test battery, 121
Test for intrafamilial conflictual
   constellations, 52
Thematic Apperception Test, 4
Therapeutic aspect, 28, 29
Therapeutic effect in the first GAI
   session, 73
Therapeutic effect of the GAI sessions,
   72
Therapeutic principle, 180
Therapeutic process, 14

Therapeutic regression, 24
Therapist
   supports, 16
   in training, 122
Therapist's activity, 69
   behavior, 38
   protection, 34
Threatened in the GAI, 71
Threatening danger, 24
   imaginal contents, 71
Three-tree-test, 52
Thumbsucking, 70
Training procedure, 28, 30
Transcript of treatment, 122
Transference, 14, 30, 131, 157, 170
Transference-countertransference
   relationship, 43
Transference relationship, 82
Transference resistances, 39
Transformations, 50
Traumatically caused phobias, 46
Trees, 26
Trial behavior, 5
Trial GAI, 49
Two-track outpatient treatment, 80
Two-track psychosomatic therapy, 91

Unburdening function of GAI, 55
Unconscious constellations, 103
Unconscious drive dynamic, 4
Unconscious fantasies, 81
"Understand" the symbolic
   connections, 28
Undesirable incidents, 44

Verbal communication, 16
Village, 50
Visiting a house, 19
Vividness of imagery, 34

Weather, 17
Wishful fantasies, 16, 21
Wishful thinking, 20